Southern Illinois Fishing

Shawnee Books

SOUTHERN ILLINOIS

FISHING

A COMPREHENSIVE GUIDE FOR ANGLERS

Colby Simms

Southern Illinois University Press
Carbondale

Southern Illinois University Press
siupress.com

First published December 2024.

Cover and frontispiece illustration: Internationally renowned fishing
pro and media personality Colby Simms lands a Southern Illinois
bass on his popular School N Shad. *CSO Team / Colby Simms
Outdoors LLC.* Blue lake with cloudy sky as a part of a nature series
by TSpider (cropped). *Shutterstock (Photo ID: 118169548).*

ISBN 978-0-8093-3962-4 (paperback)
ISBN 978-0-8093-3963-1 (ebook)
The book has been catalogued with The Library of Congress.

Printed on recycled paper ♻

SIU
Southern Illinois University System

This book is dedicated to my family, from whom I derive strength and passion, drive and dedication, love and inspiration; the people I cherish from the bottom of my heart and the depths of my soul. These earthly angels have made it possible for me to chase and accomplish my life's goals and dreams, to discover the real meaning of life, to find a happiness and peace that's indescribable. This life would be nothing without you, and it is everything because of you. From the bottom of my heart, thank you, and God bless you. I love you all.

Contents

Gallery of illustrations begin on pages 19 and 111

Acknowledgments

My parents, Ray and Gloria Simms, raised my brother Shay and me in the wilderness. Growing up on a farm in the country—and boating, fishing, hunting, hiking and camping from the time we were little kids—Mom and Dad nurtured our fascination with the woods and the waters and with creatures that called these magical places home. We spent nearly all of our time in the real world—the natural world—learning to love, to cherish, and to respect nature and its rhythms, to develop and hone senses in order to become one with it, and to stoke the fires of passion for true life, pure life, a life in the wild. Mom and Dad exposed Shay and I to the great waters of Southern Illinois as young children. As adults many years later, we're still in love with this place.

My Dad has always been my favorite fishing partner. And he and I have fished with each other more than anyone else in our lives. Dad taught me the skills that I needed to catch fish when I was barely old enough to hold a pole—with no idea that I would take it so far—that I would center most of my life around it. He taught me, and then I taught him. The older that I got, the more important fishing became to me. I learned absolutely everything that I could about the sport. I spent every spare second practicing it, branching out to new waters with new techniques and targeting new species every chance I got. And Dad shared in most of these experiences with me.

After I accomplished my lifelong dreams of becoming a fishing professional and outdoor sports journalist, living and working on the water full time, and starting my own related businesses, Dad became my championship title winning tournament fishing partner, a charter guide, an outdoor sports media personality and director of the Colby Simms Outdoors LLC professional staff. We'll both tell you that this couldn't have happened without Mom running the main office, handling much of the operations as the CSO business manager, and supporting us in every way possible. She is nothing short of a Godsend. And this meant that Dad and I have been able to spend an incredible amount of time out fishing the magnificent waters of Southern Illinois. It also meant that we were able to travel extensively, inside and outside the country. Doing magazine photo shoots and network television shows at many of the world's top destinations, we learned unique and valuable information that we've used to enjoy even greater fishing success at home. It's been a dream, but one that all started right here in the bottom of the Prairie State.

Southern Illinois Fishing

Introduction

Southern Illinois is a special place for many reasons. But for the angler, these beautiful and bountiful lands offer much more. They provide something that is difficult to describe, something unique and something wonderful, something to be treasured and something to be celebrated. For it is here in this magnificent place, this sportsmen's mecca, that uplifting and truly magical experiences are had. It is here that lovers of the great outdoors can, and so often do, develop passions that will rage inside them like wildfire for all of their lives.

I'm blessed to have been able to live my dreams, as a full-time fishing pro and outdoor sports media member, for most of my adult life. And I've also been blessed to call these my home waters, the waters of Southern Illinois. Hunting and hiking, boating and camping, rock climbing and horseback riding are just a few of the countless activities and sports practiced in this place—this place that seems as if it was created just for these purposes. But it is the great sport and ancient tradition of fishing that causes the outdoor sporting paradise that is Southern Illinois to shine brightest.

Southern Illinois fishing is truly remarkable. The tremendous variety of waters and species, the awe-inspiring wilderness areas surrounding them, and the incredible action that is had from high numbers of fish combine with excellent big-fish potential from the many trophy-caliber specimens taken from these long-famed lakes and rivers, creating the perfect storm that is the sport of fishing. We'll cover all of that in this book, with information from a broad group of top fishing pros, biologists, outdoor sports media members and other experts on fish and fishing. We'll discuss all of the region's premier waters, with site-specific information. From the basics to the most advanced cutting-edge techniques, we'll offer a battle plan for attacking the fisheries of the bottom portion of the Prairie State, which all anglers, from beginners to seasoned tournament competitors, will benefit from.

America's favorite sport fish, the largemouth bass; freshwater's greatest sport fish, the mighty muskie; hard-fighting stripers and silvery whites; feisty bluegills and tasty crappies; beautiful trout and wily walleyes; and monstrous catfish are just a few of the many species that showcase what this incredible region of the country has to offer anglers. In this book we'll get into detail about these and other fish species of Southern Illinois, as well as the science behind their behavior. We'll provide invaluable insight and a wealth of on-the-water experience, expertise proven most effective to catching them.

From hot lures to top tackle selection, power boating big waters to paddling out-of-the-way spots, from weather and water conditions to seasonal changes and fishing pressure and how these many factors influence fish behavior, requiring anglers to make adjustments to their game, we'll delve into the nuts and bolts of how it's done here. From fishing fast to fishing slow and everything in between, we'll dive headfirst into all that is fishing in the lower Land of Lincoln.

By 2019 in the United States alone, the economic impact of the fishing industry had grown to $125 billion, producing over eight hundred thousand jobs. North America's number-one participation sport is practiced by more than a billion people worldwide, and Southern Illinois anglers have made contributions to this global fishing stage. Southern Illinois angling has inspired the creation of tactics, products, and techniques that have been used by recreational anglers and amateur competitors to top professional charter guides and tournament pros across the United States, Canada, Mexico, and beyond, and we'll discuss that here.

Initially for survival and eventually for sport, fishing has been important to the residents of Southern Illinois throughout its history. The region features mile upon mile of flat farm fields and low woodlands, rolling forested hills and steep rugged mountains with vertical rock faces. Every kind of terrain one can imagine—falls down to tiny creeks and gigantic rivers, small lakes and large reservoirs, little oxbows and vast swamps, shallow farm ponds and deep flooded strip pits—all filled with fish waiting to be caught.

Some of the country's best anglers have called Southern Illinois home. Some of the most interesting personalities that a writer could ever hope to interview, or an angler could ever hope to spend a day in a boat with, ply the waters here with regularity. Rest assured, there will be no shortage of fun and funny stories and interesting and inspirational material in the following chapters.

While you're likely to find this book complementary to a bowl of warm chili, a cold beer, or a cup of hot chocolate by the fireplace or the campfire, beware. You'll no doubt discover it easy, as we have, to aggravate a boss or a spouse, at least the ones who don't fish anyway, as the waters will likely beckon too much to ignore. Let yourself sink into the magic that is Southern Illinois sportfishing. And after ingesting these pages into your mind and soul, drop everything and make time for some quality time, the quality time that is fishing in Southern Illinois.

For the purposes of this book on fishing, Southern Illinois is defined as the area of the state classified by the Illinois Department of Natural Resources Region 5, or the IDNR South Region. We'll cover all fishing within this area, including all major lakes and rivers and other waters of significance, as well as all of the sport fish species and the more popular rough fish species that anglers target with sportfishing methods.

1

Food for the Soul

Stars begin to fade in the dark night sky as light slowly edges its way over the tall, forested mountainside. A bit of fog hangs in the trees as mist rolls from the calm glasslike surface of a wilderness lake. A white-tail deer magically appears for a drink as squirrels scurry around in the leaf litter searching for tiny morsels, and a loon calls to its mate off in the distance. Upon the all-too-familiar boat deck, a few yards off of a wooded shoreline of rock and clay, just past the edge of where a small vertical rock cliff face falls down to trees and scrub and scattered boulders, I notice a miniscule bit of disturbance under the water, indicating life below.

I fire a long cast with my surface lure, careful to cause the offering to land near the bank and past the tiny ripple of water. Manipulating the rod tip, I crank the reel handle, quickly at first to get the bait moving, then slow and steady. As the lure ambles along, sputtering and gurgling, throwing just a bit of water into the air, it begs a response. Here in this magical place, entranced by the beauty and perfection that is the natural world, in the great woods, out on the mysterious water, the perfection of the moment lulls one into a state that cannot be found elsewhere, a state of peace and serenity, a state of awe and of inspiration. It is a state that causes one to experience the excitement of a child on the last day of school and, at the same time, the relaxation of an old man in a hammock on a sunny early fall day.

Along this quiet lake shore, *whooooosh*! The sputtering topwater lure is suddenly and viciously attacked from a monster lurking beneath. The lake explodes, as waves of water forcefully roll away in all directions from the once flat surface. Spray flies upward, followed immediately by the head of the great beast, causing the squirrels to run from the shoreline in fear. The great fish surges up and completely out of the lake, hammering the helpless-looking topwater lure with a bone-jarring strike, a hit that would surely knock any small creature unconscious. It snaps its powerful jaws at the lure as its beautiful and muscular body rockets into the air, like a white shark predating an Indian Ocean sea lion from the surface off Geyser Rock.

The massive fish turns in midair and crashes back down into the lake. The entire attack happens in just an instant. But the mind of a seasoned angler has been trained, conditioned to slow it all down, savoring every single bit of this unmatched experience as one savors every bite of the finest of meals. But we've also trained ourselves to slow this lightning-fast event down in order to make a perfect execution. For as all serious anglers know, timing the hookset of a predator gamefish's attack on a topwater lure is absolutely critical.

Topwaters require a hookset at just the right moment. Most types of fishing with artificial offerings allow anglers to immediately set the hook when the fish strikes, something much easier to do than when fishing the surface. Certainly, one of the most difficult acts in the great sport of angling is to watch a fish attack a lure on top and remain calm. Holding back the nearly indescribable excitement, long enough for the fish to get the lure firmly in its mouth, can be agonizing and goes against every instinct swimming around inside of us. Still, it must be accomplished.

I wait in anticipation, gut-wrenching anticipation, until I feel the weight of the fish at the end of the long line. And then when I sense it, know for certain that it's not missed the bait, know that it has in fact made a firm grip, I forcefully pull the rod tip hard, up and to the side. I rotate at the hips, turn the bulk of my body the opposite direction as my quarry, and, with a vicelike grip on the rod and reel handle, swing my shoulders back with a violent twisting motion, generating as much force as humanly possible to drive the steel home.

Upon proper completion of the technique I am rewarded, as I feel the sport fish punish me with fury while it pulls back hard with the same full-body kind of force that I just employed. Feelings of satisfaction, of accomplishment, and of exhilaration flood my mind and soul, as adrenaline burns like fire throughout my body. After a head settles into the softness of a pillow in the dark, of this is what the angler dreams. This is what we live for!

In this age-old game of cat and mouse, pursuing the sports that those who came before us spent their lifetimes to master, relishing in the ancient traditions of give-and-take with the earth that has allowed our species to survive and thrive for thousands of years, we reach a hard-to-find place within ourselves. Down deep in the wilds of our soul, we find the spot on the map of life, arrive at the destination where we were destined to go.

DAD AND I HIT THE WATER early one morning in beautiful Southern Illinois. We were primarily after largemouth bass and muskies but were up for whatever came along. The fishing had been good for a week, producing some of the biggest crappies of the season and a heck of a walleye. The action was picking up, but a storm was approaching, and a cold front was coming through that night. We awoke in the dark of early morning to a blow. The weatherman got it right, several days before, actually. The moderate winds of the three previous days had turned to high sustained winds with frequent hard gusts. While blowing out of the north a full week ago, they had gradually shifted through the previous days, now coming mostly out of the south. Today, the winds were howling hard out of the southwest. And the partly cloudy, partly sunny conditions of the previous days were replaced with dark cloudy skies. The storm was nearly upon us, and immediate prefrontal conditions were set in for the duration now.

Excited for what was to come, we hurried out the houseboat door and headed for the covered docks. Dad opened the air tank valves on the lift and dropped my 21-foot bass boat in the water. We'd had all the rods and reels rigged up and ready to go the previous day, so we wasted no time in getting to our prime spot just as it was barely light enough to see. As we motored out of the marina, I put the hammer down and slammed the foot throttle to the floor. The big 250-horsepower outboard jumped the fiberglass boat up on plane in an instant, and we screamed across the lake at over 70 miles per hour. Coming up on the first spot of the morning, I eased back on the throttle, and we slowly came down off plane. Motoring along the outside edge of the structure, sloshing around in big waves, my sonar unit confirmed what we had guessed. Tons of baitfish and gamefish were stacked up along this windblown irregular shoreline break. We were seeing bait and fish from the surface down to around ten feet. We surmised that there would be at least as many fish up in the shallows, right on top of the structural element, the prime location for active feeding fish to be on an early-morning rampage.

We backed out just a bit more so that long casts would place our lures up on top of the structure in very shallow water, right along the irregular shoreline bank and the points protruding out into deep water. This allowed us to cover the entire shallow and mid-depth structures, as well as the deep water just past the edge of the drop-off. With small flats and shelves here and there in between the points along the bank, there was plenty of excellent shallow structure to target. And this area, one that we'd been taking huge fish from for years, had been productive earlier in the week. Chunk rock and submerged patches of weeds provided ample cover for predators to hold and hunt in, from the shallows out to moderate depths. A sharp break on the edge dropped off into deeper water, providing excellent habitat for all species. It allowed gamefish to set up and remain there to feed without the need to move far. The water was moderately clear, with more than enough clarity to be able to easily see shallow subsurface lures running all the way from the bank to the boat.

I started off with a large buzzbait on top, and Dad went subsurface with a triple-bladed spinnerbait. We fished hard and aggressively, firing long casts and burning the lures at high speeds back to the boat in an attempt to cover a massive amount of water as quickly and efficiently as possible. Every couple of minutes, as the boat rocked back and forth in the big waves, the wind whistling past, a bit of spray came up into the air and over the side of the boat. Within no time, a fish exploded out of the water, engulfing my surface lure in its powerful jaws! I set the hook hard and immediately felt the weight and strength of the beast pull back hard against me. While staying on the trolling motor to keep us positioned properly to fish the structure, and to keep the boat from crashing onto the rocks, I fought the fish for what seemed like a while. It dug for the depths most of the time but made several spectacular jumps high into the air. Finally, the big bass came to the boat, pushing six pounds, as Dad reached over the side and put a lip lock on its gaping maw. After a couple of quick photos, she was back in the lake and we continued on.

Over the next hour or so, we each caught several more average-size largemouth plus a couple of small ones. Dad landed a nice one of about four pounds, and I put a big fat white bass in the boat. Within minutes of releasing the white, Dad yelled from the back deck: "I've got him!" I happened to be glancing back toward the rear of the boat just as it happened. On the passenger side of the back deck, Dad jerked the rod tip hard as he yelled. The big rod doubled, and I knew that he had a monster of a fish on. I was looking toward Dad. But he was watching his spinnerbait come through the clear water. And he got to, once again, experience the heart-stopping sight of a big muskie viciously and savagely attacking a fast-moving lure just under the surface!

The battle was on, and he went toe-to-toe, like a heavyweight fighter in a career highlight slugfest. The monster muskie pulled back hard against him, running right, then reversing its direction and going back to the left. Dad put it to him hard and gained line, pulling back on the rod tip and then reeling down quickly as he slowly moved the great fish toward the boat. I grabbed the huge net and then jumped back on the trolling motor, to ease us out into deeper and more open water. Here we could more easily land the fish without the chance of it getting wrapped up in cover, and without having to constantly worry about the boat getting smashed onto the rocks in the high wind and big waves.

The fish dug for the depths while out away from the boat but then came up and jumped twice. Like a submarine, the muskie plunged right back down and stayed deep for several seconds before rushing back up again to tail walk across the surface of the lake, shaking its big head quickly and violently back and forth in a desperate attempt to throw the lure and win its freedom. Like a true champion, the muskie gave us everything we hoped for and then some. And, also like a true champion, my dad fought the great beast, with the perfect execution he'd used to win titles in tournament competition. Even with all the other top fishing pros I've shared a boat with in my life, there hasn't been one that I would put money on over my dad for being able to land a big fish. Today would be no exception.

Dad started to come toward the front deck, but then the muskie surged again and took off from the back of the boat. He jumped back onto the rear deck, and again worked her back to the boat. Trying to go around the boat on the driver's side, he pulled her back and steered her head the other way. In a valiant last-ditch effort, she headed for the motor. We often trim the motor up when fighting a big fish, but not in big wind. The outboard acts as a rudder helping to stabilize the boat. Luckily by this time we had passed well around the end of a point, and the wind was blowing us toward open water. So, there was no need for Dad to come back to the front of the boat, as I could finally get off the trolling motor. He guided the giant fish over and around the motor and steered her big head toward the passenger side of the boat again, our favorite place to net a big fish.

Jumping down into the bottom of the boat, I leaned over the side from behind the passenger console. She shook once more, and Dad used the long rod to guide her headfirst, right toward me. In one motion, I dropped the end of the netting that I was holding in my forward hand against the net handle near the hoop, and scooped the trophy caliber fish up. Catch! And she was ours. What a beautiful barred fish this was. A good 25 pounds of slimy razor-toothed fury. But just another monster we've taken from that prime structure, when big winds are bashing it. The great beast went back into the lake unharmed, ready to fight again another day. Despite the big-fish conditions, we saw few other anglers on the lake that day. And of the ones that were there, most left early in the day. Of all the anglers we saw out fishing, every one of them was in a quiet, protected area, and we never saw anyone catching fish.

We ran into some anglers at the marina that evening, and none of them did well that day. Few fish were caught among them, and nobody took any big fish. We continued to fish the evening. Action was steady most of the time. We had two other flurries of high activity, catching not only numbers of fish but big ones too. We landed three more muskies, hooked another, and had several other follows. We lost count after catching a couple dozen more largemouth bass, including several big ones. We even popped a big channel catfish and a drum on spinnerbaits. Was it tough fishing in those conditions? Yes, it was. Was it more difficult to fish and to hold the boat on the wind-blasted structures in big waves? No question. But it paid off for us. It paid off as it so often does, when we forget about the rest of the world and become fully connected with nature, and when we take all of the conditions into account and utilize that information to nail down precise patterns.

It was another truly phenomenal day on the water here in magnificent Southern Illinois.

We are all a part of nature. Like most of the outdoor pursuits that have sustained human life, fishing is not merely an activity, not just recreation, not simply a sport. No, immersion and participation in the outdoors is a primal need inside of us all, whether we have come to accept that truth or not. It is a desire to know, with great intimacy, how to experience life in the real world, the natural world. To feel the very essence of what it means to be alive, surging through our veins. Like an old man drinking from the fountain of youth, we are rejuvenated by nature and the great sports that we pursue there. It is here, in the raw unadulterated savageness that is nature, in the calm of the water and the quiet of the forest, in contemplation of this natural world, its wild creatures, and of our relationship thereto, that once-lost souls are discovered again, that we find the fuel we need for living. And, Southern Illinois is a fine place to find yourself lost in the modern sport and ancient tradition of fishing.

2

The Fish

Southern Illinois abounds with fish species. From large lakes and reservoirs to small lakes, flooded strip pits, and farms ponds, from massive rivers to little creeks, springs and swamps and oxbows, the region's available aquatic habitat is diverse. The area offers countless opportunities for different styles of fishing, for many species of gamefish, and more.

Black Bass

America's most popular fish, largemouth bass are some of the most sporting quarry and the primary black bass species available to Southern Illinois anglers. They're a real challenge to pursue and can be difficult to figure out at times. But when conditions are good, and we nail down a precise pattern of where they are located and what they're willing to bite on, the action can be fast and furious. It isn't uncommon to catch 50 to 75 bass per day on the region's top bass waters in spring and fall, and we've had days over the years in which just a pair of us have landed over 100 bass from a single boat.

The size of these fish is also impressive. The better bass lakes here consistently produce lots of quality largemouth, from 3 to 5 pounds or better. For the latitude and growing season, it's one of the best areas in the country for coughing up big bass from 6 to 8 pounds or so. Fish over this weight class aren't common anywhere in this part of the world, but there are more caught here than many others. Incredibly, a number of giants from the 9- to 12-pound class have come from several different Southern Illinois waters!

While largemouth are certainly caught on all kinds of live baits, they take every kind of artificial lure imaginable at one time or another, including topwaters. And of course, this adds to their popularity, since more anglers prefer the use of lures over live bait, feeling that it's a more exciting way to fish. What's more is that artificial lures are usually the best option, allowing anglers to fish faster and cover more water to contact more bass.

Largemouths, or bucketmouths as they're often referred to, are formidable. While they can school up together to roam a body of water in search of prey, most of the larger specimens are primarily ambush predators. They conserve energy while concealing themselves, waiting on prey to carelessly amble into their strike zone. They're known for vicious, heart-pounding strikes that often feel as if they'll rip the rod from your hand, and they'll give you a slugfest battle. They dig for the depths one second, then leap high into the air the next, in a beautiful acrobatic display of strength and will. Yes, bass fishing is a roller-coaster thrill ride at a mile a minute!

Smallmouth bass are a fan favorite of anglers all over North America, and for good reason. They strike and fight very hard for their size. While they don't grow nearly as big as largemouth bass, many anglers believe that they fight harder pound for pound. They're beautiful creatures. Smallies are often called bronzebacks, with a kind of golden-brown color. They can have shades of green and gray too. Spotted bass also fall into the family of black basses here. Spots look similar to those of largemouth bass, but these have smaller jaws. Similar in size to the smallmouth, Southern Illinois spotted bass often school up and can provide fast action. Like all bass, spots also fight hard. They're a lot of fun to catch on light- to medium-power tackle.

Crappie

Crappie would have to take the number-two spot for popularity on the list of favorites for SI anglers. While most bass enthusiasts pursue the species entirely for sport and release all of their catch, the opposite is true for crappie fishers. Most hit the water with the intent to put a limit of fish in the cooler or live well to take home, and for good reason. Crappie, also called specs, are some of the tastiest freshwater fish you'll ever eat, with white flaky flesh that has a flavor far better than most.

Specs fall into a category with bluegills and sunfish, known as panfish, since keeper-sized fish usually end up in a frying pan. Like other panfish, all three species of crappie, white, black, and hybrids, are also more prolific. They reproduce quickly and are found in higher numbers than many other species. So, taking fish for the table isn't as likely to affect the fishery in the same way. While white crappies are longer than blacks, black crappies are wider than whites. White crappies tolerate turbid water much better than black crappies do, for anglers who prefer to target one species over the other. Crappies tend to school together in large groups, and anglers often catch numerous fish from one spot before moving on.

Live-bait fishing is probably about as equal in popularity as using lures for crappie. Southern Illinois has many excellent crappie waters that cough up high numbers of fish. And Southern Illinois is home to two of the three crappie state records, both well over 4 pounds. It's a little more difficult to rank the remaining species by popularity. Some of these fish are very popular with locals but draw little fishing pressure from tourists. Others species, however, draw more attention from traveling anglers than from locals.

Muskie

Muskies are one of these species. The number of local anglers that routinely fish for muskies in Southern Illinois is small when compared to bass and crappie, or even catfish. But Southern Illinois muskie fishing draws large numbers of traveling anglers, including many Canadians. My pro staffers and I have guided charter trips for anglers from most American states and Canadian provinces, and even for anglers coming from as far away as South America and Europe, since Southern Illinois's muskie fishery is truly world-class.

In defense of local muskie angling, there is a much smaller number of lakes in the region that harbor muskies than ones that don't, and the number of local muskie anglers is impressive considering the fishable waters. Muskie fisheries have been created by the state and require regular stocking to maintain populations of these cold-water northern fish. The number of stocked fish per acre of water varies considerably from lake to lake based on a number of factors. While fish from around 3 to 6 inches in length are heavily preyed upon, most muskies are stocked at larger sizes. Fish from about 10 to 14 inches have much higher mortality rates.

Muskie fishing popularity is skyrocketing, both in Southern Illinois and across much of North America. These are massive fish, big and powerful, fast and strong. A number of Southern Illinois muskies in the upper 40- to low 50-inch range, from 30 to 37 pounds, have come from these famed waters. However, it's the fast action that truly sets the region apart. When compared to muskie fishing in most waters, in both the United States and Canada, Southern Illinois waters yield hotter action and faster fishing than most. Like they do with bass, or even more so, almost all anglers targeting muskies fish them exclusively for sport. Even so, minimum size limits placed by the state are so high that all but the biggest specimens must be released. Still, when anglers catch legal-sized fish, they're almost never kept. And those who practice the sport of muskie fishing not only consider release a cornerstone of the activity but also carefully follow cutting-edge release techniques to further increase muskie survival.

While occasional fish are taken with very lively live-bait offerings, artificial lures are always the best way to catch these fish. Anglers primarily cast and retrieve big lures on heavy tackle, but jigging and power trolling can be effective during certain times of year as well. Muskies are some of the toughest customers swimming the world's freshwaters. Pound for pound they strike and fight even harder than bass do. And with their impressive size, hooking up is a level of challenge and excitement that's difficult to match in freshwater.

In addition to raw striking and pulling power, they race around the boat, often changing depth and direction during a fight, and they jump frequently. They're definitely more thrilling than most. The muskie is considered to be the greatest of all of North America's freshwater gamefish and a pinnacle for angler accomplishment, so much so that the National Freshwater Fishing Hall of Fame in Hayward, Wisconsin, is highlighted by its 4-plus-story-tall, 143-foot-long walkthrough muskie museum sculpture, bigger than a Boeing 757.

Temperate Bass

All the temperate bass species are found in Southern Illinois. The most common are white bass, simply called whites. Yellow bass and white yellow bass hybrids are present. Striped bass are the largest, commonly called stripers. The hybrid cross between the striper and white bass, hybrid striped bass, are often referred to as wipers. These fish are often released, but some people keep them for the table as well, and they're good eating whitefish. These tough guys are most known for their fighting abilities. Powerful creatures, they strike and fight very hard. But it's the length at which these battles go on that really sets them apart.

While these fish don't typically jump as often as black bass and don't race around quite like muskies do, they just never seem to give up. And many anglers feel they pull harder pound for pound than any other fish in freshwater. They vary widely in size and numbers. In good conditions anglers can catch well over 100 whites in a day, with bigger fish running 2-1/2 to 3 pounds. Yellow bass and hybrid yellow bass are smaller, but they can produce feverish action as well. Stripers and wipers are caught in much lower numbers. But big Southern Illinois hybrids can run into the teens, with the former state record an SI giant over 20 pounds. Pure striped bass on the other hand routinely grow 15 to 25 pounds and can surpass 30 here.

Sunfish

Bluegills and their close cousins, the various species of sunfish like redear, longear, pumpkinseed and green sunfish, warmouth, and rock bass, are what most of us start off with as children. There's nothing better for getting kids hooked on fishing. Even though they're the smallest gamefish in size, with most specimens weighing well under a pound, they fight very hard for their size. Using light and ultralight tackle makes catching them great fun. What makes them so great for kids is their tendency to bite. In almost all waters, there are far more of these fish per acre than any other gamefish species.

They compete with one another more for prey and are naturally more aggressive than most gamefish. Live bait is more popular, but a variety of small lures work well on light line. When conditions are right, it's possible for anglers to actually catch hundreds of these feisty fish per day on numerous waters in Southern Illinois and still have a chance at a trophy gill or sunnie over a pound. Southern Illinois is the proud home of both the state record bluegill and the state record redear sunfish, and Illinois is one of just seven states in the entire country that has produced a bluegill exceeding 3-1/2 pounds!

Trout

Rainbow trout are available in Southern Illinois in a limited number of waters. These fish are also entirely dependent on stocking to maintain their numbers, and fish are stocked either once or twice per year depending on the location. Good fishing can be had at times for bows up to around 6 pounds or more, and these feisty and beautiful creatures can sometimes produce fast action. A species of salmonid, the rainbow trout is a cold-water fish that is most commonly taken on small offerings. The light tackle commonly used to target these fast-swimming gamefish means an exciting fight every time.

Walleye and Sauger

Walleye are another fish found in limited numbers in the region, and one also primarily dependent on stocking. While catch rate numbers are typically low, some big-trophy-caliber fish have been taken over the years in the 7-to-10-pound range. A close cousin of the walleye but considerably smaller in size and preferring turbid waters, the sauger also makes Southern Illinois its home. This fish thrives best in lower Prairie State rivers. The saugeye is a hybrid cross between the two species, making an exciting catch for anglers here.

Catfish

Southern Illinois catfishing can be incredible. These very hearty and adaptable creatures can be found everywhere. Channel catfish are the most pursued quarry, found in excellent numbers in most SI lakes and rivers. Catching large numbers of channels from 5 to 15 pounds is common, with giants from the 30- to 40-pound class possible. Monstrous blue catfish swim many Southern Illinois waters, especially the larger rivers. Southern Illinois produced the former state record blue. The 85-pound beast came out of a deep hole in the Mississippi River in 2000. This state record held until it was bested by the new 58-inch-long world-record 124-pound behemoth blue cat also from the Mississippi River, but further north near St. Louis. Trophies from 40 to 70 pounds are common, especially in the winter months when big blues congregate in deep holes of the main river channels. They're a fun fight, especially in heavy river current.

Big flathead catfish are also caught throughout much of Southern Illinois in both rivers and lakes. The flathead is a top predator that fights incredibly hard and stays down deep, digging for the bottom during battle like few freshwater fish do. A number of area waters have produced trophies in the 40-to-70-pound range and even bigger, but 20- and 30-pound brutes are common. The much smaller bullheads are fun on light tackle, and are often caught while fishing for both channel catfish and panfish.

Rough Fish

I need to mention rough fish. They don't get much attention from anglers. They're often called "trash fish" and are seldom bragged about. But native rough fish species play important roles in the ecosystem and help keep things in balance for the more popular gamefish species. There are a number of various species in Southern Illinois that don't fall within the gamefish category, and this is the same in most states. This is interesting too, because most of these fish grow to large, and in some cases gargantuan, proportions. Most fight incredibly hard when hooked and put up a far better tussle than some gamefish species do. Some are challenging to find and catch, while others are fairly easy and provide fast action. But all of them are a lot of fun and worth the chase from time to time.

Common carp are transplants from Europe and Asia, where they are considered the most prized sport fish in many countries. Believe it or not, in much of Europe they view carp like we view bass. In Illinois, these tough fighting fish can surpass 50 pounds. Grass carp grow even larger, over 70 pounds here. While most rough fish are caught with live or dead natural baits, the freshwater drum is more likely to attack an artificial lure, like top predator gamefish do. A cousin of the highly prized saltwater red drum, or redfish, freshwater drum commonly grow to over 30 pounds in Southern Illinois, including the massive 35-pound state record.

Spotted, shortnose, and longnose gar are fun to catch and will also take artificial lures in addition to natural baits. The biggest of these, the longnose, can grow to over 20 pounds here, with the Ohio River producing the state record at over 22. Southern Illinois's second state-record gar was a spot that went 7 pounds and 13.44 ounces at Horseshoe Lake.

The giant alligator gar is set to make a comeback in Illinois, with a recent reintroduction program by the Department of Natural Resources, to include stocking into the lower Kaskaskia River. In recent years, just across the river from Southern Illinois, Missouri's biggest alligator, a 2007 bow-fishing record that stands at 127 pounds, was taken in the headwater diversion channel of the Mississippi River in Cape Girardeau County. Black, smallmouth, and bigmouth buffalo, herring, goldeye, various shad and sucker species, and other rough fish can be found swimming these waters. And Southern Illinois is home to both of the 16-pound, 6-ounce bowfin to tie for the state record, coming from Bay Creek and Rend Lake.

Multiple championship title winning tournament pro Colby Simms lands an SI bass on his CST Hatchet Spin. *CSO Team / Colby Simms Outdoors LLC*

CSO pro Ray Simms caught this monster muskie on a Hatchet Shad in the bottom of the Prairie State. *CSO Team / Colby Simms Outdoors LLC*

The author's friend BWA TV host Mark Davis displays a stringer of nice crappies. *Big Water Adventures / DL Ventures LLC*

A magnificent wintertime sunset over a beautiful Southern Illinois wilderness lake. *CSO Team / Colby Simms Outdoors LLC*

CSO Team's Amber Ronketto Simms landed this big nighttime SI flathead catfish. *CSO Team / Colby Simms Outdoors LLC*

Colby Simms hooked into this huge SI walleye on a School N Shad while giving an interview for MJO Radio. *CSO Team / Colby Simms Outdoors LLC*

A hard-fighting bass leaps from Southern Illinois waters trying to shake a School N Shad spinnerbait. *CSO Team / Colby Simms Outdoors LLC*

CSO Team guide Walt Krause with a School N Shad and a big Southern Illinois striper. *CSO Team / Colby Simms Outdoors LLC*

Ultralight spinning equipment makes fishing for bluegills an absolute blast. *CSO Team / Colby Simms Outdoors LLC*

CSO Team's Chris Shannon with another big Southern Illinois bass taken on the author's Thump N Shad. *CSO Team / Colby Simms Outdoors LLC*

Amazing waterfalls abound in prime fishing areas of the Lower Land of Lincoln. *CSO Team / Colby Simms Outdoors LLC*

CSO Youth Team's Cade Simms caught his first bass with his grandmother, CSO business manager Gloria Simms. *CSO Team / Colby Simms Outdoors LLC*

Championship title-winning tournament pro Ray Simms works the figure eight on a following muskie. *CSO Team / Colby Simms Outdoors LLC*

Record-holding fishing guide Colby Simms lands a hard-fighting Prairie State white bass. *CSO Team / Colby Simms Outdoors LLC*

CSO Team's Chris Shannon shot this photo of sons Nathan and Nolan Shannon enjoying a day of trout fishing. *CSO Team / Colby Simms Outdoors LLC*

CSO Team's Amber Ronketto Simms caught this trophy SI largemouth on a Hatchet Spin. *CSO Team / Colby Simms Outdoors LLC*

Bass are Southern Illinois's and America's favorite fish, and choosing the proper tools to chase them is critical. *CSO Team / Colby Simms Outdoors LLC*

CSO Team's Chris Shannon was fishing with the author when he caught this giant SI muskie on a School N Shad. *CSO Team / Colby Simms Outdoors LLC*

Hard-fighting blue and channel catfish are abundant in many Southern Illinois waters. *CSO Team / Colby Simms Outdoors LLC*

Sport fishing pro and public speaker Colby Simms landed this trophy Illinois crappie on his CST Hatchet Spin. *CSO Team / Colby Simms Outdoors LLC*

3

The Fishing

Southern Illinois has and continues to hold a special place in the heart of anglers who experience it. I've spent many years now basically getting paid to be on vacation, traveling and fishing extensively throughout the United States and other countries. As incredible as it has been to take in the very best of many of the planet's top fishing destinations, I still always enjoy coming back to my home waters of Southern Illinois. It's a true gem in the world of sportfishing.

State within a State

For those of you who, like me, are residents in the land of the Shawnee National Forest, we are incredibly fortunate. At 286,000 pristine acres, the magnificent Shawnee is the largest publicly owned body of land in the state, and often seemingly overflowing with wildlife. For those of you who travel to this unique place from elsewhere, you know the appeal to return as often as possible. Both groups however, yearn to soak up as much information as is available about the waters, the fish, and the fishing methods that produce results with consistency, to experience all that this region offers the angler.

Most of the lower portion of the state does not resemble the rest of Illinois. Endless flat stretches of prairie and cropland turn to rolling hills and wetland swamps, steep gorges, and tall rocky peaks of the Southern Illinois Ozark Mountains, lying between the Ohio River, for which the Buckeye state was named, and America's largest river, the mighty Mississippi. Fishing opportunities here rival top regions all over North America. The sheer variety of waters and species available makes this a very special place. But add to that the fact that numerous truly world-class fisheries exist here, and it's easy to see why many locals rarely travel outside of the area to practice their favorite sport, and why anglers from all over the nation and even other countries journey here just for the fishing.

Southern Illinois abounds with fish and fishing styles. In this central latitude within the continent, we experience great opportunities for both warm-water species found far to the south and cold-water species of northern fish. The methods that anglers use to catch our various finned friends are just as varied. Fishing with both artificial lures and natural baits happens frequently for many different species on all waters. The use of bait-casting tackle and spinning tackle is most common here, but some anglers are dedicated to using fly tackle as well. From shallow water and power-fishing tactics to deep water and finesse techniques, SI anglers use it all to score lots of big fish in the Lower Midwest.

Casting methods reign supreme most of the year for most species, on the majority of waters, and can be done by boat or on foot. Vertical jigging is also highly effective at times, usually requiring the use of a boat. Still fishing is very popular for specific fish, and especially for shore-bound anglers, while power-trolling tactics garner plenty of attention from boaters, especially those plying the larger lakes and reservoirs. Some specialized fishing styles and tricks used in Southern Illinois are not practiced much outside the region, while some of North America's now hottest tactics and techniques were developed right here.

Southern Illinois has always been a popular region for tournament fishing as well. Many local fishing clubs as well as state and regional tournament circuits hold events on the big waters like Crab Orchard, Egypt, Kinkaid and Rend Lakes, and the Ohio and Mississippi Rivers. But many small waters host events as well. Additionally, a number of large tournaments of national-level pro circuits have been held here. And since Southern Illinois is centrally located between a number of other very popular areas of the country for recreational fishing and high-level tournament competition, the region becomes a good central location for the traveling competitive angler. Regardless of just what makes up your particular cup of fishing tea, you can find it in the lower Land of Lincoln.

While my love for the sport began in Eastern Missouri where I grew up, it was in Southern Illinois where my passion for fishing really took over my life and guided me down the path to my dreams. Even when living in Missouri we were just a short drive away. And my family and friends and I considered our home waters to be in Southern Illinois, where we made more trips to fish than anywhere else. After I took up the sport full-time, I moved here and began fishing Southern Illinois waters over 250 days a year. And with a childlike love and fascination, I could never get enough. Southern Illinois fishing fostered inside me a passion for this sport unlike few things that exist in this world.

For a number of years, I still spent tremendous time fishing here. But I traveled a lot and fished many other places too as my tournament career progressed. The experiences were priceless and helped me to catch even more fish in Southern Illinois. But these were some very busy seasons. After years of touring across North and Central America, competing in tournaments for muskies, bass, and billfish, and appearing in many television programs, I was ready to mostly retire from that part of the job. I wanted to slow things down a bit and spend more time back home with family and friends. Tournament fishing really is a lot of fun. I relished the competition and looked forward to each event. But at the same time, it was stressful too. There was a feeling of relief at the end of each event, and especially at the end of each season. Even though I won numerous tournaments, took two championship titles, and had a number of other high finishes, records, and awards in various circuits, it was always a difficult way to attempt to earn a living.

I've had some decent sponsorships over the years. But good ones are hard to come by, and it's certainly not what most people think. With the anxiety of putting enough fish in the boat for TV film crews, and the anxiety of putting enough fish in the boat during competition to win enough money to pay some bills, while juggling charters, giving seminars, pushing products, and working trade and sport shows, it can be taxing. Still, I'm glad that I did it. I mostly look back on the experience with fond memories. For the last several years now, I've continued guiding charters on my home waters here in Southern Illinois while writing and running my outdoor sports businesses.

My good friend Andrew Adank would be a charter guide's dream client. He reminds me of the pure joys of fishing like when we were all kids. As an adult, and especially after years of full-time pro angling, I used to sometimes get a little frustrated when the bite was off and fishing was slow. In all the great many years we've fished together, mostly in Southern Illinois, I can say that Andrew thoroughly enjoys being in the boat more than most, regardless of what the fish are doing. We've had some incredible days on the water, catching many big fish together. But he is just as content to catch an occasional small fish, or even none at all. And it's a refreshing change of pace from the folks who take it too seriously.

I'll admit that I took it all too seriously for a number of years. Especially when I was traveling all the time and doing a lot of television and tournament fishing, when each bite counted so much. But as time has passed, I've become more laid-back, stopping more frequently to smell the roses. Or maybe more appropriately here in Southern Illinois, to stop and smell the honeysuckle. And thoroughly taking in the experience of just being on the water. In the end, I think it actually makes you a better angler too. Simply being content in the great outdoors helps one achieve a state of deep relaxation, and a stronger connection to the wild, without all of life's stresses. The sport of fishing has never come so easily, so naturally, as when I'm in a good mood. And it's these times in which I've had most of the greatest fishing days of my life.

The Tackle Industry

On the waters of Southern Illinois, I've designed and tested various lures and other fishing and hunting products for many different companies that I've worked with over the years. Back in 2003, I developed my own line of lures and terminal tackle items here as well, Colby Simms Tackle. Applicable for a wide range of species, our muskie products and our bass products have produced countless trophy-caliber fish for regional, national, and international network TV programs and magazine article photos. Over the years, we've produced and offered about 600 different SKUs of items, through retail dealers throughout the United States, Canada, and Mexico, including the world's largest tackle retailer, Bass Pro Shops.

Though tested in fresh and salt water throughout many countries on multiple continents, all of the designs were created for Southern Illinois species, on this region's great waters. Southern Illinois has a long tradition of importance to the fishing tackle industry, with several other companies founded and operated here that offer products nationally or internationally. Southern Illinois University alums Chris Piha and Matt Gunkel operate Llungen Lures of Carbondale, originally founded by long-time Illinois and Minnesota guide Chad Cain, producing a wide range of lures mostly for muskies and pike. Lunker Lure was founded decades ago in Carterville before joining with Hawg Caller and moving to Du Quoin, offering a range of items, primarily for bass fishing.

Unfortunately, with overseas manufacturing and giant corporate conglomerate domination of the industry, the business has changed, and many smaller operations have scaled back or fallen out. Two Southern Illinois tackle businesses no longer in operation were once important to the lure industry and Southern Illinois fishing. SI Crappie Baits offered a variety of excellent panfish lures. Dunn's Big Bite Lures, a former sponsor of mine, sold a huge variety of bass fishing products through their chain of Dunn's Sporting Goods stores in Missouri, Illinois, and Kentucky. Over the years there have also been a number of local SI anglers that handcrafted their own custom lures and tackle items that they primarily sold locally, which were responsible for some big catches too. All of the SI-born products have made Southern Illinois an even more important location on the international sportfishing stage.

For a number of years, our primary business focus was our lure company. But the main focus now is on our outfitting operation. We refer our friends and clients and book trips for fishing and other outdoor sports charters and lodging, with our partners at dozens of our favorite fishing destinations throughout North and Central America. I still travel, just not as much as I did in the past. I enjoy the simplicity. I enjoy getting to once again spend the vast majority of my days fishing my home waters here in Southern Illinois. Waters that I know so well and love so much. Yes, this place is dear to my heart. And while I'll always enjoy traveling and fishing in many different places around the world, I'll always enjoy wetting a line here on these magnificent waters.

I live near the water when I'm in the Show Me State. My house and office sit along a fork of Plattin Creek, just across the drive from my parents' home near the lake on our family farm. Still, I'd always dreamed of living on a houseboat in Southern Illinois and being able to fish 24 hours every day on my favorite waters. After winning a first-place victory in the Professional Musky Tournament Trail event at Cave Run Lake in eastern Kentucky, and qualifying for the PMTT World Championship on Lake Minnetonka in Minnesota many years ago, I used prize money and bought my first of two houseboats where I've lived about half the year since.

For an angler, there is nothing better than living on a houseboat. In essence, you reside where the fish do. Large houseboats have all the comforts of a modern home on land but allow anglers to fish constantly. They can be docked in the marina, tied up along wilderness shorelines, or anchored just about anywhere on the water. Houseboating offers a lifestyle and a fishing experience unlike anything else. And a number of my friends and clients are now planning to live aboard somewhere when they retire.

Seldom Considered

Smart anglers utilize every method to learn more about fish and their waters. I've had a number of good fishing spots on various lakes and rivers over the years that I regularly chum. Chumming keeps large numbers of fish using these locations, and not only can we catch these fish, but we can study them too. Anyone can do the same with a chum dispenser. Thick plastic jars and bottles work well. Drill a bunch of holes in the bottom and sides and drop in a rock for weight. Fill the container with scented fish chum and squirt in a little crayfish or baitfish oil attractant. Then attach it to something above or near the water by cord, drop it down in the water column, and leave it there to work its magic.

When you go fishing, swing by and drop a little cheap dog food or stale bread in the water first. Then toss some live and cut up shad or minnows after the feeding gets going. You can attract and catch almost all species. Some catfish anglers utilize chumming methods. But outside of catfishing, few anglers pursuing other species chum at all. For those who have the time, it definitely pays off. Even for anglers after fish like bass, crappies, and muskies.

In some places, they get so used to us we're actually able to feed them by hand. Some species come up to the boat like dogs and practically beg. Sunfish, catfish, and even rough fish like carp will literally mob us trying to get food. While bass are a little more wary, we have been able to feed them by hand as well, even big ones. And, probably the most exciting thing of all is the rare yet heart-pounding occasion when you are feeding and a muskie rushes out of nowhere to grab a bluegill that's caught up in the frenzy! I've had big muskies slam into me as they attempt to grab a smaller fish. Sometimes they miss but sometimes they catch one, and it takes your breath away every time. Polarized glasses help you see further into the water. Usually, we don't see them until the strike. But if you pay close attention, once in a while you'll notice one hanging out under the boat, under a log, by a rock, or in some thick weeds nearby, like a wolf watching a herd of elk.

In the cold months, we usually just drop the food in the water. But it doesn't take long for them to get used to eating right from your hand again come early spring. As soon as it warms up, we jump in the water and swim with the fish we feed. If you don't have a wet suit on, it's best to wear a shirt. We've found out the hard way that apparently a bluegill's favorite thing to eat is a nipple! It's also best to keep any moles covered, or at least be aware of where the fish are because we've had them actually tear a mole right off of us. The most painful situation that we run into at times are aggressive catfish. You have to keep an eye on them and be prepared to swim or push them away.

Catfish have what are referred to as horns, serrated spikes that fins attach to. These have a film on them that some people call poison. And anyone who has ever been horned by a cat knows that this pain almost cannot be described. Getting horned is rare. But what's not rare is a catfish tearing up your hand as they take food from you. They often bite down hard and quickly shake, ripping your flesh with the rough patches in their powerful jaws. Most of the time you can feed them without getting chomped on yourself. But it's when you're out of food and getting ready to climb back in the boat or up on shore that they'll get you. For this reason, I usually leave a big chunk of food until the end. I hold it out away from my body and then let it go, so they can fight over it as I get out.

We routinely get cut up when landing, unhooking and releasing muskies, and the result is something far different than bass thumb. Besides mouths filled with big razor-sharp and needle-pointed teeth, they have sharp gill rakers too. When holding a muskie for pictures, we slide a hand under the belly to support the fish's weight and prevent damage. But the only way to securely hold one is by placing the other hand inside the gill plate, sliding it up to the point, and gripping tightly. But you have to be careful when you do so as not to grab the gills themselves, which can hurt the fish and cut your hands. We've actually had muskies bite us in the net, and unhooking fish can always be a challenge. Still, it's all worth it. The very best days of an angler's life are when their hands look like they went through a meat grinder, and you're on the cusp of needing a blood transfusion. Simply awesome!

Still, as much damage as we've taken from muskies in the net, we've never had one bite us underwater when we're feeding. But it's not impossible. Swim goggles work for seeing what you're doing, but I usually don my dive mask and snorkel to feed fish so I can keep my head underwater and not have to come up for air. Beyond feeding though, we love to slap on a pair of fins and snorkel or scuba dive with the fish. You can really learn a lot about their behavior while in the water with them. And you can discover structures and cover objects that hold fish, to come back and target later. Obviously, the lakes and streams in Southern Illinois with clearer water are best for this. But you can see okay in waters with a light stain in the shallows, with the right sunlight. Regardless of the water though, a dive light is great to have when getting down in the 30-to-60-foot range.

After all these years, I can say that snorkeling and scuba diving is the only thing that I enjoy as much as fishing. Fish are naturally curious and often show little fear of a human when that human is in their world. Even big fish will often swim right up to you. This is interesting, to say the least, when it's a big muskie or a monster flathead. I haven't seen big stripers while diving yet, although I'm hoping to. I haven't seen big gar either, but I've talked to other divers who have, and the situation sounds much like a muskie encounter. Once, while I was snorkeling the shallows of a clear-water creek arm, a muskie of at least 30 pounds swam up close to me and just hovered, wiggling its fins as if it were preparing to attack!

I just stared into those eyes, wondering what the culmination of thousands of years of predatory instinct could be. Muskies have attacked swimmers before. It's rare, but it does happen. And like in the ocean, most accidental attacks by creatures from muskies to mako sharks happen when the water is murky and the animal mistakes a human for something else. The muskie didn't choose to strike, but her eyes just kind of told me that she was sizing me up, that her brain was saying, *Can I eat that?* I came back and fished that spot that evening. I never saw her again, but I did catch a smaller male. I knew it was a good spot, and years later I landed a fat 47-incher right where she approached me. Was it the same muskie? Probably not, but big fish spots remain big fish spots.

Throughout childhood, Mom and Dad always loaded the boat up and took my brother and me to the lakes and rivers often. I snorkeled a lot with Dad and Shay as a kid, and got even more serious about it as an adult. I got into deeper free diving, and eventually got certified through PADI and started scuba diving too. Still, snorkeling is one of the simplest and most enjoyable experiences there is, and one that most anglers can do. It allows you to become one with the fish, another creature sharing their world. You observe them, learn about them and their behavior, interact with them, and come to understand their environment more than you could just by fishing.

I introduced my wife, SIU alum Amber Ronketto Simms, and my nephew, Cade Simms, to this great tradition, a natural sport for just about any angler. And it's become a passion for both of them. We fish together and we dive together, and each of the two sports compliments the other. Southern Illinois is a mecca for anglers, a place covered with waters that are as beautiful as they are bountiful. Like boating and like chumming, snorkeling and scuba diving are practices that help anglers catch more fish while allowing them to develop a better understanding of the fish, their behavior, the locations they use and why. All invaluable tools in an angler's box.

It's important to note that throughout this book, we'll be discussing things like weather conditions, fish location, and fishing presentations that all combine to make up what we call patterns, the ways to catch fish on a given day. To avoid too much repetition, some important pieces of information might appear in just one or two sections of the book, like when I'm talking about walleye angling techniques or the behavior of catfish. Some anglers, who primarily just fish for bass and crappie for instance, might be tempted to dismiss information presented about, say, trout or gar. But this is a mistake. From my lifetime of fishing and many years of full-time professional angling, as well as fishing with countless other pro anglers, biologists, and dedicated recreational fishers as an outdoor sports journalist, I can say that one of the most important things an angler can learn as they practice this sport is a willingness to be versatile and to not dismiss anything related to fishing in any place or for any species. It all crosses over. And the best anglers across the planet embrace this mindset.

4

The Waters

We'll discuss most of Southern Illinois's fishable waters in this chapter. But we'll focus on the more popular lakes and rivers, the waters with the best fisheries, and those of special significance to the region at the outset.

Big Muddy River

The Big Muddy River is a 156-mile-long waterway draining Franklin, Jackson, Jefferson, Marion, Perry, Union, Washington, and Williamson Counties. It is dammed to form Rend Lake and spills into the Mississippi just south of Grand Tower, Illinois. The most popular places to fish are in the tailwater area below Rend Lake dam and the lower section of the river from around Murphysboro to its confluence with the Mississippi. But many species can be found along its full length. Illinois Department of Natural Resources fisheries biologist Jana Hirst lists the Big Muddy as one of her top picks for largemouth and spotted bass, white and black crappie, redear sunfish, channel, blue and flathead catfish, and yellow and white bass.

Catfishing is rated as excellent. Live sunfish, goldfish, and shad are top baits for Big Muddy flatheads most of the year, while cut shad are the best bet for blues. I've shared a boat with Bill Gottschalk, owner of Harrison's Sport Shop in Cambria, who considers the Big Muddy one of his favorites for channel catfish. Bill usually prefers stink baits and chicken livers for this species. Blade baits, as well as jig and grub combos, are some of the better lures for black and temperate bass, which rate good. Crankbaits and blade baits produce drum, gar, and bowfin, and so do live shad and minnows. Big carp are suckers for night crawlers on the Big Muddy, which will produce channel cats, gar, and other species too.

Cache River

The Cache River is a 92-mile-long waterway. Its basin spans Alexander, Johnson, Massac, Pope, Pulaski, and Union Counties, and the area is part of the largest complex of wetlands in the state of Illinois. The Cache drains into both the Ohio and Mississippi Rivers between the areas of Mound City and Willard. Channel catfishing is rated as excellent, and the river produces good fishing for largemouth bass, crappie, and bluegill. Flathead and yellow bullhead catfish are present. Some blue cats can be caught as well, especially in the lower section. Many rough fish are available, like carp, drum, buffalo, bowfin, and various gar, including the rare spotted gar. The grass pickerel, another rare species to the region, can be found here.

Jana Hirst lists the Cache as one of her top choices for black and white crappie, largemouth, spotted bass, redear, channel cats, white bass, and pickerel. Night crawlers, smaller live and cut shad, and big live minnows are top baits for the abundant channel cats, and minnows are also the top bait choice for crappies. Spinnerbaits and jigs produce largemouth, spotted bass, and whites all year round, and frogs take largemouth in the warm half of the year. Spinnerbaits are best for pickerel and gar, and jigs with mealworms produce lots of redear, bluegill, and other sunfish.

Cedar Lake

Cedar Lake lies in Jackson County near the college city of Carbondale. Its 1,750 acres of water are partially surrounded by Shawnee National Forest, including many acres of pristine wilderness areas complete with tall mountainous hills, boulders, and rock bluffs plunging into Cedar's beautiful waters. This reservoir has a maximum depth of 60 feet and provides a wealth of angling options. Largemouth bass fishing on Cedar is rated as excellent. It's definitely one of Southern Illinois's best bets for anglers in search of a trophy-sized bucketmouth. Cedar has given up largemouth over 8 pounds in recent years. Fishing is rated as excellent for redear sunfish, and also for both black and white crappie, with high concentrations of large-sized specimens. This lake offers bluegills and other sunfish and produces fair to average fishing for channel catfish as well as a few big flatheads, all underutilized species. Cedar is also one of the few Illinois lakes offering striped bass, and a number of big linesides from the 25-to-30-pound class have been caught in recent years. IDNR fisheries biologist Shawn Hirst, Jana's husband, lists Cedar as his top pick for stripers and largemouth bass, his third choice for black and white crappie, and his top pick for bluegill and redear.

One of our original CSO pro staffers, my good friend Walt Krause, has been guiding charters on Cedar for many years and has put our clients on some of the biggest stripers that Illinois has produced. Top options include large diving crankbaits, multiblade spinnerbaits, and minnow baits. Buzzbaits, spinnerbaits, and crankbaits are top choices for largemouth bass. Suspending jerkbaits can really shine in the cool months, and small soft plastic worms and lizards get the nod when the bite is tough. Medium shiner minnows produce crappies best when the bite is slow, but tubes and small shads work best most of the year on jigs. Drop-shot fishing with these same artificial lures or live minnows is also becoming a productive pattern here, especially when the fish are deep. Hirst advises that while predators will eat anything they can catch, most favor gizzard shad on Cedar.

Centralia Lake

Located in Marion County, Centralia Lake is 254 surface acres of water southwest of Salem. With a maximum depth of 23 feet, this body of water has nearly 13 miles of shoreline. Largemouth bass offer the lake's best angling opportunity and provide great numbers of quality-sized fish, although white crappie fishing is nearly as good. The fishing for bluegills, as well as channel and yellow bullhead catfish, is fair here, and the lake also harbors redear sunfish and some large carp. Long Texas rigged ribbon or curly-tail plastic worms and 5- to 6-inch freak baits are top options for bass here. Anglers catch high numbers of crappie on marabou jigs or soft plastic tube jigs, as well as live minnows. Big carp take scented dough baits here the best, but worms produce bruisers as well. IDNR fisheries biologist Boone LaHood advises gizzard shad and bluegills to be the most important prey species here.

Crab Orchard Lake

Located in the remarkable 43,890-acre Crab Orchard National Wildlife Refuge in Williamson County, near the towns of Carterville and

Crainville, Crab Orchard Lake is one of Southern Illinois's largest bodies of water at 6,965 surface acres with 125 miles of shoreline. Considering its large expanse, it is a relatively shallow body of water. Crab Orchard Lake, or simply Crab as most locals refer to it, has a maximum depth of only 25 feet and an average depth of just 9 feet. The fishing has been steadily improving on Crab for a number of years now for multiple species. Channel catfish angling is rated as excellent on Crab, and flathead catfishing is good too. While bluegill fishing is only average, crappie fishing is rated as excellent, with black and white crappie available, along with some other sunfish. This reservoir has good largemouth bass fishing that keeps getting better, to include the nearly 12-pound mammoth bucketmouth landed during a tournament just a few years back. Crab Orchard Lake offers very good white bass fishing. Large carp are present, and the lake produces larger than average bowfin as well. Camping is available in this wilderness paradise.

Shallow-diving squarebill crankbaits, spinnerbaits, and bladed jigs are top options for active largemouth bass, while flippin' jigs and ribbon-tail worms get the nod when horizontal presentations aren't working. Buzzbaits and tailspinner prop baits take quality fish in the warmer half of the year too. Blade baits, wide wobbling casting spoons, and lipless cranks are usually best for white bass. Smaller blade baits also take crappies. But underspin jigs score many of the biggest Crab specs with baby shad and shiner-style soft plastic bodies. Large live minnows and big pieces of cut shad are most often the best choice for big channel cats, but stink baits often produce numbers. IDNR fisheries biologist Luke Nelson advises the most important prey species here to be gizzard shad, sunfishes, and aquatic invertebrates. Nelson lists Crab as his top pick for largemouth and white bass, white and hybrid crappie, channel and flathead catfish, carp, and drum, and his second choice for bowfin and gar.

While powerboaters are able to fish more water than others, many lakes and rivers in Southern Illinois have great places to fish from the bank. When it comes to the bigger lakes and reservoirs, Crab has some of the best shore fishing opportunities around. Fishing the highway banks and bridges on foot has always been a popular and productive pastime at Crab. The lake also features long rocky fishing piers that jut far out into the water, which many anglers take advantage of. Some have easy drive-to accesses that are utilized by many handicapped anglers as well. These are surrounded by underwater fish attractors that hold lots of baitfish and gamefish, adding high quality cover to draw more prey and predators to the structures. Anglers fish from the land in these and other places on Crab, but the use of waders, belly boats, canoes, and kayaks are becoming more popular here and elsewhere too.

Devil's Kitchen Lake

In the Crab Orchard National Wildlife Refuge, between Wolf Creek and Makanda in Williamson and Union Counties, lies Devil's Kitchen Lake. Kitchen, as we locals refer to it, is 810 acres of deep water with about 24 miles of shoreline. This long skinny body of dammed water plunges to a depth of 90 feet and provides great and varied fishing opportunities. Devil's Kitchen produces good fishing for largemouth bass and bluegills and very good fishing for redear sunfish. Catfish and carp are available, as are crappie, warmouth, and other sunfishes. Trout fishing is what the Kitchen is really known for. With a significant amount of very deep water and cooler temperatures in summer, the lake maintains a very good year-round trout fishery. And as a result, it only requires stocking once a year, in the fall. It also offers a developing yellow perch fishery, a unique species to Southern Illinois. Camping is available, and hiking and hunting are popular in the forested areas around the lake. Luke Nelson favors Kitchen as his top choice for rainbow trout, his second pick for bluegill and redear, and third for black crappie. He shares that sunfish, spotted sucker, gizzard shad, and aquatic invertebrates are top prey items.

In the cooler months of the year, Devil's Kitchen trout are suckers for shallow-diving suspending minnow plugs. But deep-diving versions take fish from the depths in warmer months. Marabou and bucktail hair jigs, inline spinners, and spoons are great here too. One of the best bites of the year however, especially for big bruiser rainbows in the heat of summer, comes from fishing live night crawlers on deep bottom structure, especially at night. A big plus is that this tactic produces big carp, channel cats, bass, and various sunfishes at the same time. Although it works better if an angler chums the area first. Crankbaits and spoons are top choices for active bass, but drop shotting straight-tail worms and other finesse plastics works great for less-active fish. Ned rigs have also become a hot ticket for bass at the Kitchen in recent years. Maybe the best option is a one-half or one-third section of a stick worm on a mushroom-head finesse jig. Yellow perch, redears, bluegills, and other sunfish are best here on live baits. Minnows, crickets, mealworms and red wigglers are top options.

Du Quoin Reservoir

Du Quoin City Lake is north of Du Quoin in Perry County and is a real sleeper for panfish. This is a 210-acre lake, with 12 miles of shoreline and a maximum depth to 30 feet. The Du Quoin reservoir provides excellent crappie fishing. Catfish angling has dropped off and remains poor, but good fishing for largemouth bass and bluegills exists and just keeps getting better. Redear sunfish angling is also excellent here. In fact, this body of water probably provides the best opportunity for redear sunfish in the entire region. Du Quoin is also home to the state record freshwater drum, a beast that weighed in at 35 pounds! Largemouth bass are usually best on small swimbaits and spinnerbaits. Soft craws are a top choice too, especially when the fish are less aggressive. Crankbaits and blade baits are always top choices for big drum, but live minnows and night crawlers work great as well. Bluegills and Du Quoin's many trophy redears love mealworms, wax worms, crickets, and grasshoppers, but these fish can be taken on most soft plastic insect or minnow imitators fished on small jigs. Jig spinners work well too. Crappie fishing is hot here on tubes and marabou jigs, but live minnows are tough to beat if the bite slows.

East Fork Lake

East Fork Lake is located just north of Olney in Richland County. At 935 surface acres it reaches a maximum depth of 40 feet and an average depth of 15 feet. East Fork boasts good fishing for walleye, largemouth bass, channel catfish, and crappie. The waterbody also offers bluegills, redear sunfish, yellow bullheads, and some large carp. Boone LaHood advises gizzard shad and bluegills to be the most important forage items here, and lists East Fork as his top pick for black crappie and walleye. Eyes are taken here on jig and grub combos, spoons, and crankbaits. But live baits can produce better when the bite is off. Large minnows and night crawlers produce walleyes, a bonus being that these baits are top choices for the lake's channel and bullhead catfish. Jigs paired with soft craws or chunk trailers take some of the bigger bass here, as do crankbaits. Drop shotting baby shad and shiner plastics leads to some of the bigger crappie catches, but switching to medium shiners on the same rig can produce better at times.

Ferne Clyffe Lake

While small, Ferne Clyffe Lake makes the list as one of Southern Illinois's top trout fisheries, with some of the best scenery and hiking as well, and camping available. Located south of Goreville in Johnson County, Ferne Clyffe is just 16 acres of water in the Ferne Clyffe State Park. Fishing for largemouth bass and redear sunfish is rated as very good here. There are also crappies, bluegills, and good numbers of quality-sized channel catfish that are typically fat and heavy bodied. But the main draw here is the lake's quality rainbow trout fishery that rates very good. Trout are stocked here in spring and fall, and hiking and climbing the beautiful trails and magnificent rock formations ices the cake on a trip to this unique place. A great way to fish this small lake is for multiple species at the same time, with live baits. Night crawlers, mealworms, red wigglers, crickets, and minnows can all take rainbow trout, channel catfish, largemouth bass, bluegills, and redear. Trout also take flies, inline spinners, marabou jigs, and scented trout dough baits. Bass hit soft plastic worms and craws well, and the panfish are fond of tube jigs. Catfish also take dough baits. Luke Nelson lists only sunfishes and aquatic invertebrates as prey species of significant importance here.

Forbes Lake

Located in Marion County about 13 miles northeast of Salem, Forbes Lake is 585 surface acres of water with a maximum depth of 31 feet and about 20 miles of fishable shoreline, with camping available. Forbes has a very diverse fishery, with some unique species. The largemouth bass fishing here rates very good and has produced some big specimens over the years, the crown jewel of the lake's angling opportunities. White crappie, bluegills, and redear sunfish are available. Forbes Lake produces good fishing for both channel catfish and yellow bullhead catfish as well as some very large carp. Forbes provides average fishing for saugeye, and offers rainbow trout and hybrid striped bass. Most of the hybrids caught here, saugeyes and wipers, are taken while fishing for bass or catfish. Top options for bigger largemouth include crankbaits, and paddletail swimbaits or curly-tail grubs fished on plain jigs and underspins, all of which can produce wipers and saugeyes too. Live medium shiner minnows are a top choice for channel and yellow bullhead catfish, which also produce bass and hybrids. Boone LaHood lists Forbes as his top pick for wipers and advises gizzard shad and bluegills to be the most important prey species.

Kinkaid Lake

In Jackson County, between Ava and Murphysboro, lie the 2,750 acres of Kinkaid Lake. Situated within the beautiful Southern Illinois Ozark Mountains of the Shawnee National Forest, the reservoir is surrounded by nearly 10,000 acres of wilderness lands. This large and popular impoundment has an average depth of 25 feet and offers 82 miles of shoreline. While one end of the lake is somewhat flat, with fields and gently sloping wooded hills, most of the water is surrounded by steep hillsides and tall rugged mountain tops, with many vertical rock bluff walls and giant boulders. Kinkaid is one of the deep waters of Southern Illinois with a maximum depth of 80 feet, and it offers a very diverse fishery. Muskie fishing is rated as excellent. It's widely considered one of the world's premier fisheries, often producing high numbers, and has given up monsters surpassing 35 pounds. Another world-class option, crappie fishing also rates excellent here, with lots of quality fish and Kinkaid producing the state record hybrid, tipping the scales at 4 pounds and 8.8 ounces. White bass are fair, but there's very good largemouth bass fishing and a few smallmouths. Bluegill, redear, warmouth, pumpkinseed, and green sunfish are available, along with large carp and drum. Its channel cat fishery is good, and there are a few flatheads and bullhead catfish. Kinkaid also offers an improving walleye fishery, with some very big specimens. Camping is available, and boating, hiking and hunting are popular.

Kinkaid is Shawn Hirst's top pick for muskie and walleye, his second choice for largemouth bass and black and white crappie, and his top pick for hybrid crappie. He considers gizzard shad the most important prey species on Kinkaid. Muskie action is best on spinnerbaits, crankbaits, minnow baits and topwaters. Jerkbaits and swim jigs produce when the bite slows. Bigger largemouth bass are often best on swimbaits, spinnerbaits, and crankbaits, but soft plastic worms, beavers, and small freakbaits can produce faster action from smaller fish. Kinkaid's crappie bite happens best on tubes and soft shads fished on jigs and underspins, but medium shiners and fathead minnows produce better on a slow bite. Small spinnerbaits and crankbaits also produce big specs in warmer months. Walleyes and white bass are best on crankbaits, spoons, and blade baits, but minnows produce too. The most popular baits for channel cats are molded stink baits and punch baits. But the garlic-salted, strawberry Kool Aid–soaked chicken craze has reached Kinkaid, and many swear by this strange combo. Sunfishes are best on mealworms and red wigglers but take hair and tinsel jigs well too.

Lake of Egypt

Lake of Egypt's 2,300 acres of water lie south of Marion in Johnson County between Goreville and Creal Springs. This body of water has a maximum depth of 52 feet and an average depth of 18 feet, boasting 93 miles of shoreline. A campground is available, part of the Shawnee National Forest, but this is one of the few major lakes in Southern Illinois with many private homes along its shores. LOE produces good fishing for largemouth bass and crappie. Bluegills and channel catfish are available, and the lake produced the former state record hybrid striper of over 20 pounds.

Topwater fishing is productive for bass here. Poppers and other chugging-style lures are a hot ticket, but buzzbaits and frogs produce quality fish too. Punching vegetation with plastic beavers, worms, and craws also takes many of the lake's big largemouth. Crappies are best here on tube jigs, marabou underspin jigs, or live minnows, while catfishing shines on shrimp, night crawlers and stink baits.

Little Grassy Lake

In the Crab Orchard National Wildlife Refuge of Williamson County, between Makanda and Wolf Creek, you'll find Little Grassy Lake. This clear body of water is over 1,000 acres in size and plummets to depths of around 60 feet. Little Grassy produces fishing for largemouth bass, bluegills and redear sunfish that rates as very good for all, and it also offers channel catfish. Little Grassy Creek downstream from Little Grassy Lake spillway is a noted whitewater kayaking location that provides up to class 4 whitewater during significant water flows. Camping is available here. Luke Nelson advises that gizzard shad, sunfishes, spotted sucker, and aquatic invertebrates are all important prey species. He lists Little Grassy as his top pick for black crappie, yellow bass, and pickerel, his second choice for green sunfish, warmouth, and flathead catfish, and his third pick for largemouth bass, white crappie, bluegill, redear, and channel cats.

Plastic tubes, marabou jigs, and jig spinners are top choices for bluegills and redear sunfish here when fish are active, but crickets, mealworms, and wax worms produce better when the bite is slow. An inline spinner can be productive for these fish too. Frogs, buzzbaits, and walking-stick baits are all hot for bass in the warm months of the year, while spinnerbaits, jigs, and plastic worms take largemouth all year round. Numbers of catfish are caught on dip and punch baits as well as hot dogs.

Mermet State Lake

Not to be confused with the popular private scuba diving destination Mermet Springs just to the north, Mermet State Lake is located about 10 miles north of Metropolis, near Joppa, in the Mermet State Fish and Wildlife Area Park of Massac County near the Ohio River. A shallow lake, Mermet is 452 acres of water with a maximum depth of just 12 feet. Redear sunfish and bluegill angling is rated as excellent here. The lake offers average channel catfishing, as well as very good fishing for largemouth bass and crappie. Hunting and hiking are very popular here, and the site is the home of Illinois's Pro-Am National Archery Tournament. Big redears and gills are taken here on jigs, flies, and crankbaits, but red wigglers and mealworms are hot live baits. Crappies are best on minnows. Bass take stick worms, soft jerkbaits, and spinnerbaits readily. Luke Nelson advises Mermet to be his top pick for bluegill, warmouth, redear, and green sunfish, as well as bowfin and gar, and his second pick for white crappie and channel cats.

Mississippi River

Nicknamed the Father of Waters, the Big Muddy, and the Mighty Miss, the great Mississippi River is the world's fifth largest river by volume. At 2,202 miles, it comes in a close second in length in the States only to the Missouri River. While locks and dams turn much of the upper portion of the Prairie State's stretch into a series of reservoirs, Illinois's lower section from Alton to Cairo closely resembles much of the lower Mississippi, which runs from the confluence of the Ohio River on the Illinois/Kentucky border to the Louisiana delta in the Gulf of Mexico.

It's characterized mostly as unimpeded, fast-moving open river, with strong main channel currents. The section from Alton to Cairo, Illinois, offers some 61,266 acres of water to fish. In Illinois's Southern Region, the Mississippi winds its way along the state border the length of Jackson, Union, and Alexander Counties from north to south, to its confluence with the Ohio.

Catfish are king in the Mississippi. Fishing is rated as excellent for blue catfish, flathead catfish, and channel catfish throughout this stretch of river. High catch rates are the norm with many huge specimens available, truly world-class angling. Sauger are common in the main river, and walleye are available. Fishing for temperate bass is also very good in the main river too. White bass are the most common, with large numbers of quality-sized fish caught. The Mississippi produces some very large striped bass and hybrid striped bass. Grass pickerel, largemouth bass, smallmouth bass, spotted bass, black and white crappie, bluegill, green and redear sunfish, and warmouth are available, especially outside the main river areas. Large pallid, shovelnose, and lake sturgeon can be caught and released as well.

The Mississippi is loaded with large rough fish. Common carp, grass carp, smallmouth and bigmouth buffalo, drum, bowfin, American eel, longnose gar, shortnose gar, spotted gar, and alligator gar can all be taken with sportfishing methods. Other large fish exist here, invasive rough fish species like bighead and silver carp, and a native sport fish, the paddlefish, also called spoonbill. However, since these are filter feeders that siphon plankton from the water much like whales, they will not strike lures or baits presented with traditional sportfishing methods and must be snagged, usually by ripping weighted treble hooks through the water during the season. Various Mississippi tributaries are very important to a number of species, and Jana Hirst lists spotted bass and white bass in particular among them. Mississippi tributaries contain most of the primary forage fish commonly found in the main river as well.

When it comes to catfish, Dad and I have spent more time chasing these species on the Mississippi than most other waters, and their trophy potential is unmatched in the region. Blue catfish always bite best on cut baitfish pieces. Sometimes whole dead fish will produce, especially smaller ones, but it's still best to make a couple of cuts into the fish so that the blood and oils of the flesh can create more scent. Sections of skipjack herring, gizzard shad, threadfin shad, and mooneye are the top blue cat baits here, but cut pieces of various species of chubs, suckers, shiners and other oily baitfish produce well too. Mississippi River flatheads prefer big live baits most of the time, and the livelier the better. Hardy baitfish that live long on a hook in current, such as any of the various sunfishes, suckers, and chubs, will work, but smaller rough fish like carp and drum produce too. Lively shad, herring, and mooneye produce flats, but these must be changed more often. Flatheads will take cut baits when they're less active such as in winter and after the spawn, and the oily fish work best due to increased scent. But day in and day out, it's live active prey that gets the big ones.

Channel catfish will take the same cut baits that produce blues and all of the live fish that attract flatheads, but usually in smaller sizes. Night crawlers are just as good a bait for Mississippi channels as fish are, however, and sometimes they prefer a gob of several big crawlers. Shrimp, grub worms, crawfish, and molded stink baits produce too. Temperate bass take rattling crankbaits and lipless cranks, blade baits, rattling spoons, spinnerbaits, and paddletail swimbaits or curly-tail grubs fished on underspin jigs. They also bite many of the lively baitfish species mentioned for catfish. And drum, bowfin, and gar will take anything that produces temperate bass. These are also suckers for night crawlers, which produce gigantic carp better than any other bait.

As Dad would always say, the river requires respect. Excluding the oceans, the Mississippi is probably the most dangerous water I've ever fished, especially the section from St. Louis to the Gulf of Mexico. Boat choice here is about safety first and fishing second. A craft should be large enough and stable enough for the swift waters, with at least two people being able to stand and fish off the same side of the vessel at one time. Motors should be regularly maintained and running smooth, but it's a great idea to have a small kicker outboard, or at least a powerful electric trolling motor in case of a breakdown. With rock dikes or wing dams and other submerged hazards, large boat traffic, tugboats with barges running the river, and barges tied along the shoreline, just floating downstream with no control could be fatal.

As a teenager I ran my wide-bottom 16-foot-long johnboat on the river with a 35-horsepower outboard. But I never took my narrow-bottom 14-foot johnboat on the river with its 10-horsepower outboard. In fact, the 16-footer is probably the smallest craft I'd venture out onto the river with even today. Experience is also important. Anyone considering starting to fish the Mississippi River by boat should go out with someone else first, to learn. The first time you come out of a slack area into the main river current flow and feel the power of that river grab, lift, and move your boat around under power, you cannot help but feel nervous. The river takes hold of you, and it takes time to learn how to safely navigate it.

Luckily, shore fishing can be extremely productive as well. Anglers can drive a car to the more popular fishing spots or hike or horseback ride into more remote areas. Some places can also be reached by truck or jeep, with or without oversized tires depending on the spot. ATV fishing is becoming more popular too. Anglers use 4-wheelers and side-by-sides to haul big-river fishing gear down to the water's edge. We enjoy putting the trucks and SUVs in 4-wheel drive and off-roading to remote fishing spots along the Mighty Miss. We usually have a choice of several big soft sandy beaches all to ourselves, where we like to build big bonfires with driftwood. We drive rod holders into the ground and set out lines with live or dead baits for catfish, carp, drum, and others, and we'll wade fish the sandbars and cast lures for whites, stripes, bass, gar, and various species.

Murphysboro Lake

Murphysboro Lake is located within the Murphysboro Lake State Park, just west of the town of Murphysboro in Jackson County. Its 145 acres reach a depth of 36 feet with 8 miles of shoreline. Surrounded by tall, forested hills popular with hikers, this scenic lake produces excellent fishing for largemouth bass and channel catfish. There are bluegills and pumpkinseeds, and it offers good fishing for crappie and very good redear angling. Camping is available. Shawn Hirst lists it as his second top choice for redear and channel catfish, and third for largemouth bass. Hirst advises that gizzard shad are the dominant prey species.

Janet and her late husband Terry Graeff, who owned and operated Top of the Hill Bait Shop in Murphysboro for many years, fished it often and turned me onto the hot redear sunfish action there. While I had mostly targeted crappie and bass at Murphy Lake, I was pleasantly surprised by its redear fishing. The Shawnee Muskie Hunters Club holds its annual kids fishing derby there every June. The kids target various species, including these feisty panfish on light tackle. Sunfishes eat hair and tinsel jigs and mealworms best. Live shiner and fathead minnows are the top bait choice for quality-sized catfish and crappies here, and they will take most other species too. Murphysboro bass are best on jigs with small freakbait or beaver trailers, as well as ribbontail worms. Buzz frogs also produce big specimens in the warmer months of the year.

Newton Lake

Newton Lake can be found in Jasper County 6 miles southwest of the city of Newton. Newton Lake is 1,750 acres of water with a maximum depth of 40 feet and an average depth of 16 feet. It has 52 miles of shoreline, and hunting and hiking are popular. Newton is a cooling lake with an electric power plant operating on its shores. The hot water discharges keep much of the lake considerably warmer in the cold months of the year, and it's become a popular wintertime destination, especially for bass anglers. Panfish populations here are below average, but bluegills and black crappies are available. Channel catfish angling is rated as excellent on Newton Lake, and there are many large carp. Largemouth bass fishing and white bass fishing are both very good, but the largemouth bass get the most attention with high numbers of quality bass and a great shot at a trophy. Boone LaHood advises that the main forage species are gizzard shad and threadfin shad, and he lists Newton as his top pick for largemouth and white bass.

My good friend Mark Davis is a longtime fishing pro, journalist, and industry insider. He was director of marketing for some of the sport's biggest and oldest fishing product companies before going on to host his own award-winning, national saltwater fishing TV show on the Outdoor Channel, World Fishing Network, and My Outdoor TV, *Big Water Adventures with Mark Davis*. Mark and I have fished together and done outdoor sports media work at some of the world's best destinations for freshwater and saltwater angling. He turned me onto Newton Lake many years ago, when we fished with guide Tab Walker and stayed at his Outdoor Sportsman's Lodge. Since then, Newton has continued to give up numbers of big SI bass, offering probably the best wintertime fishery in the entire region. Like me, Mark and Tab love a good topwater bass bite. Because of the amount of hot water discharged by the power plant, anglers amazingly catch bass routinely on buzzbaits, prop baits and topwater cigar plugs here in winter. Spinnerbaits, tubes, and jerk worms also produce largemouth well, while cranks and blade baits are better for white bass. Channel catfish are also suckers for blade bait here, especially in summer, but will take minnows, shad, various worms, and all kinds of stink baits all year long in the warmer waters.

Ohio River

The 981-mile-long Ohio River begins in Pennsylvania. It runs 133 miles along the border of Illinois and Kentucky, in Gallatin, Hardin, Pope, Massac, Pulaski, and Alexander Counties from north to south, beginning at the mouth of the Wabash River northeast of Old Shawneetown, down to the Ohio's mouth on the Mississippi River at Cairo. In Illinois, the Ohio is divided into three river pools created by lock-and-dam systems and one section of open river south of these to the Mississippi. The Ohio offers tremendous angling opportunities from a wide variety of fish species. The river produces very good largemouth bass and spotted bass fishing, along with a few smallmouths too. For temperate species, the Ohio offers anglers very good fishing for white bass, striped bass and hybrid striped bass as well.

Good fishing for sauger is available, as well as white and black crappie, bluegill, and other various sunfish species. Channel, flathead, and blue catfish are all found in abundance, as are big carp, drum, buffalo, bowfin, and multiple species of gar. This includes big longnose gar, like the 22-pound, 1-ounce state record longnose pulled from the Ohio, in Massac County. Jana Hirst lists the Ohio as one of the region's top picks for river largemouth and spotted bass, as well as for smallmouth bass and walleye, though these are fewer in number. It's also one of her top picks for sauger, black and white crappie, redear, blue, channel and flathead catfish, stripers, wipers, white and yellow bass, and pickerel. She advises that various tributaries are important and productive fisheries as well. Jana lists gizzard and threadfin shad and skipjack herring as top prey species.

Various live and dead minnows, shad and other baitfish, night crawlers, crankbaits, and spinnerbaits produce the Ohio's record class gar. Largemouth, smallmouth and spotted bass all take spinnerbaits, spoons, bladed jigs, and crankbaits best when they're active. For black basses in a neutral or negative feeding mode, plain jigheads and underspin jigs score consistently with curly-tail worms, grubs, and paddletail swimbait trailers. All the temperate bass species here will take these same lures used for black bass, and both blacks and temperate bass are caught on live shiner and fathead minnows, shad, and other lively baitfish too. Flathead catfish favor live sunfish, shad, chubs, and carp, while blues bite cut herring, shiners and shad the best. Channels here take all of these in smaller sizes, plus various worms and stink baits. Sauger hit live shiner and fathead minnows best but will strike a spoon, blade bait, or jig-and-grub combo too.

Pyramid Lakes

Pyramid State Park in Perry County is south of Pinckneyville, offering ten managed strip pit lakes. Blue Goose Lake is the smallest at just under 20 acres, Blue Wing is 35 acres, Boulder has 25 acres of water, Canvasback comes in at 60, Goldeneye is a 125-acre lake, Green Wing is 50 surface acres, Mallard is about 75, Merganser is just over 20, Red Head is 75 acres, and the largest body of water in the system, Super Lake, is a 230-acre fishery.

Fishing rates as good to excellent for multiple species on various lakes. Largemouth bass and crappie fishing is most popular, but channel catfish, bluegills, redears and other various sunfish species are available. The most unique aspect of the waters of this park would be the muskies, walleyes, and hybrid stripers stocked there. Top patterns vary from water to water, but some of the best choices for largemouth bass are jigs, craws and plastic worms, spinnerbaits, and crankbaits, as well as buzzbaits in the warmer months. Crappie bite best on jigs with tubes or soft shads and on live minnows, while bluegills and sunfish favor crickets and mealworms on most waters. Minnows also catch plenty of catfish, but night crawlers and stink baits do as well. Muskies favor spinnerbaits and crankbaits when active but hit jerkbaits and spoons better on tough days. Walleyes and wipers both take jigs with grubs or paddletails, spoons and blade baits, as well as live minnows.

Rend Lake

Southern Illinois's biggest body of water and the second largest impoundment in the state, Rend Lake's 18,900 acres span Jefferson and Franklin Counties between Mount Vernon and Benton with the damming of the Big Muddy River. A U.S. Army Corps of Engineers flood control reservoir, this fishery has a 35-foot maximum depth and a 10-foot mean depth. Rend is about 3 miles wide and 13 miles long with 162 miles of shoreline at normal pool. Camping is available. Rend has a very diverse fishery, with over 30 species available. As far as lakes go, this is one of the best in the state for catfish. Channel cat fishing is rated as excellent, with large numbers of quality-sized fish. Flathead catfish angling also rates excellent here, producing many giant specimens. Crappie fishing rates very good, and the lake produced the state record black, tipping the scales at 4 pounds and 8 ounces. Fishing for green sunfish and warmouth is excellent here. White bass rate very good, and hybrid striper fishing is fair. Yellow bass are also available, producing the state record white hybrid yellow bass, weighing in at 2 pounds and 0.96 ounces. The lake is also home to high numbers of rough fish like common carp, buffalo, gar, and drum, including large fish of each species, and Rend produced a state record bowfin that weighed 16 pounds and 6 ounces. Largemouth bass fishing on Rend is rated as average, but is improving.

Rend Lake flatheads will eat just about any other fish that can fit in their massive jaws, including a lot of shad. But they favor the sunfishes on this water. Bluegills, redears, and pumpkinseeds are always on the menu, but green sunfish and warmouth are the top producing baits here. Channels will also take live and dead sunfishes and shad, but a lot are caught on live minnows, as well as worms, crayfish, and stink baits. Some anglers swear by the garlic salt and strawberry Kool Aid–soaked chicken for numbers of eating-sized channels here too. White bass, yellow bass and stripers take jigs and underspins with grubs and paddletails, as well as lipless cranks, blade baits and inline spinners. Crappies are best on minnows, but fat tubes and plastic critters produce on jigs. Rend's record-class bowfin can be caught on lures, but minnows and cutbait are top options. Big carp are best on strawberry, vanilla, and citrus carp nuggets. Top choices for largemouth bass include lipless cranks, spinnerbaits, freakbaits, and lizards. Shawn Hirst advises that gizzard shad are the most important prey species on Rend, and he lists it as his top lake pick for flatheads and channel catfish, white and yellow bass, hybrid stripers, black and white crappie, green sunfish and warmouth, gar, drum, and bowfin, noting that both the lake and the spillway tailwater areas are good for many species.

Wabash River

Originating in Ohio, the Wabash River flows over 500 miles southwest through Indiana, forming the border with the Prairie State until it drains into the Ohio northeast of Shawneetown. One of the largest free-flowing rivers east of the Mississippi, the Wabash runs unimpeded for over 400 miles to the Ohio. The Illinois portion is over 200 miles, and in the Southern Region it runs through Crawford, Lawrence, Wabash, White, and Gallatin Counties. Largemouth bass rate very good, and smallmouth and spotted bass are available. White bass, stripers, and hybrids produce exciting fishing. Catfishing is very good too, for large blues, channels, and flatheads, and the river produces black and white crappie, bluegill and other sunfish species, and quality sauger fishing. Large carp, gar, and drum are common as well.

Jana Hirst lists the Wabash as her top pick for smallmouth bass and a top choice for all species of temperate bass, as well as channel, blue and flathead catfish, white and black crappie, and sauger, along with a few walleyes. She advises skipjack herring and gizzard and threadfin shad to be important prey fish species. Black basses are best on spinnerbaits, crankbaits, tubes, and jig-and-craw combos, while temperate species take blade baits, spoons, crankbaits, and jig-and-grub combos best. Blue catfishing is optimal on cut shad and chub pieces, while flatheads prefer live suckers, chubs, and sunfish. Channels take both of these baits, as well as worms and molded stink baits. Shrimp is also a popular channel cat bait used here. Sauger and walleyes hit jig-and-grub combos and live shiner and fathead minnows best, and crappies are hot on marabou jigs and underspins.

Other Waters

There are other waters of significance to anglers fishing Southern Illinois. Arrowhead Lake, a small 30-acre fishery near Johnston City in Williamson County, is known for its bluegill and channel catfishing, as well as yellow bass. Arrowhead also maintains good populations of largemouth bass and redear sunfish. The lake is partially surrounded by forest and has camping available. Northwest of Glendale in Pope County, Bay Creek Lake is 123 acres of water with a depth of 15 feet. Bay Creek is surrounded by significant timbered land and produces good fishing for largemouth bass, bluegills, and channel catfish.

Beall Woods Lake is just a 14-acre body of water located in Beall Woods State Park in Wabash County, but it is one of the Southern Illinois trout fisheries, stocked with rainbow trout in spring and fall. It also offers good fishing for largemouth bass, channel cats, and bluegills and offers camping. Borah Lake lies a short distance north of Olney, in Richland County. Its 137 acres of water reach a depth of 26 feet with 15 miles of shoreline, and it produces very good fishing for redear sunfish and channel catfish. Borah also offers good fishing for largemouth bass, crappies, and bluegills, and it has a few flatheads and bullhead catfish too. Buchner Reservoir is southeast of Buchner in Franklin County. Its 17 acres offer good fishing for redear sunfish, bluegill, and channel catfish, as well as fair fishing for largemouth bass. Carbondale Reservoir is 135 acres of water located in Jackson County, at Evergreen Park in Carbondale. The lake produces average fishing for largemouth bass and channel catfish but is one of the best bets for carp fishing, with a very healthy common carp population and many large fish. In fact, Shawn Hirst lists it as his top pick for carp.

Christopher Old City Lake is north of Christopher in Franklin County and surrounded by woods. Its 20 acres produce quality channel catfish, as well as numbers of smaller bluegills, but offers excellent yellow bass fishing from a large population that anglers are encouraged to keep. Clear Creek is a Mississippi River tributary beginning in Jackson County and running through Union and Alexander Counties. Jana Hirst lists it as a top choice for spotted bass and white bass. It also produces good fishing for channel and flathead catfish, drum, and gar. Crawford County Ponds are a group of nine small bodies of water, up to 2 acres in size, located in the Crawford County State Fish and Wildlife Area. Southwest of Hutsonville, these waters have largemouth bass, bluegill, redear sunfish, and channel catfish populations and are stocked with trout each fall.

Southeast of Lanesboro in Hamilton County lies 75-acre Dolan Lake. Dolan boasts excellent largemouth bass and channel cat fishing, as well as impressive opportunities for crappie and bluegill, with water depths to 18 feet. Camping is also available. West of Dongola in Union County, anglers will find the Dongola City Reservoir, a 58-acre body of water that produces excellent largemouth bass fishing and good numbers of quality channel catfish. Dutchman Lake is a 118-acre water north of Vienna in Johnson County. Dutchman is a clear forest lake with a large watershed. It's known for its largemouth bass but offers good bluegill and crappie fisheries as well, and Luke Nelson lists it as his second top lake pick for largemouth bass and black crappie.

The Embarras River begins in east-central Illinois and runs south and east through the Southern Region Counties of Jasper, Crawford, and Lawrence before emptying into the Wabash River near Billett, southeast of Lawrenceville. Channel and flathead catfish angling is good here, and so is fishing for white bass. Largemouth, smallmouth, and spotted bass are available. Crappie, bluegill and various sunfish species, including the less common longear sunfish, are caught in this stream. Common carp, grass carp and drum are common catches too, and the Embarras River produced the state record goldeye at 2 pounds and 1 ounce. Jana Hirst lists it as a top pick for river largemouth, spotted bass, white crappie, redear, white bass, and channel cats. Hunting is popular here, in the Embarras River Bottoms State Habitat Area as well. Fairgrounds Pond, or Massac Pond, is located within the Fort Massac State Park along the Ohio River at Metropolis in Massac County. This small 3-acre lake is known for its trout fishing, with rainbows stocked annually in spring and in fall.

About one mile to the east of Galatia in the wooded hills of Saline County, anglers will find the 209-acre Harrisburg New City Reservoir. The 30-foot-deep Harrisburg Reservoir is rated as an excellent channel cat fishery and also offers impressive angling for largemouth bass and bluegills, and recent stockings of hybrid stripers could lead to an exciting future option. Herrin Lake One is a 60-acre impoundment in the town of Herrin, in Williamson County. While a major fish kill occurred in 2010, the bass fishery is rebounding, and the crappie fishing is good. Anglers targeting bluegills and redears, however, can expect very good fishing year-round. Nine miles south of town, Herrin Lake Two produces good largemouth bass, bluegill, and catfish angling. Herrin Lake Two is a 58-acre body of water surrounded by Williamson County Forest and is Luke Nelson's third top pick for flathead catfish.

Horseshoe Lake is located in the Horseshoe Lake State Fish and Wildlife Area, south of Olive Branch in Alexander County. This very large and very shallow swampy-looking lake is 1,890 acres with a maximum depth of about 6 feet. While the lake offers good crappie fishing and average bluegill fishing, the interesting future option here could be the world's largest gar. This is one of the few locations in the state where the IDNR has begun an alligator gar stocking program, which began in 2018 with adult-sized fish, a prime candidate, also producing a state-record spotted gar. Jones State Lake, also called Glen O Jones, is located in Mountain Township, southeast of Equality in the Saline County Fish and Wildlife area. The 105-acre, 33-foot-deep lake produces excellent fishing for bluegills and channel catfish, good largemouth bass and black crappie angling, and camping and fishing piers. Nearby Jones State Lake Pond is a 2-acre body of water that is known for its trout fishery. Trout are stocked in the fall at Jones Pond.

Kinkaid Creek is located near the towns of Ava and Murphysboro, in Jackson County. Kinkaid Creek has been dammed to create Kinkaid Lake. Lower Kinkaid Creek, below the Kinkaid Lake spillway, begins with a wide and deep scour hole and narrows as it winds its way several miles down to its confluence with the Big Muddy River. The tailrace area and the lower creek down to the Big Muddy are hotspots for white bass fishing. Largemouth bass, white and black crappie, bluegill, redear and green sunfish, and warmouth are all caught here, as are channel catfish. Carp, gar, drum, and other rough fish species are common too. Muskies were once caught here with regularity, but the last few years these catches have been few and far between, a result of fish staying in the lake due to a spillway retention barrier designed to keep muskies and other large fish in the lake. Upper Kinkaid Creek above the highway 151 bridge is small and shallow but does produce quality fish when water levels are normal or high.

Just west of Kinmundy, Illinois, in Marion County sits the 107-acre Kinmundy Reservoir. This lake is 26 feet deep with an average depth of 12 feet and just over 3 miles of shoreline. Kinmundy has largemouth bass, black crappie, and bluegill fishing. The lake also has a good channel catfish fishery and some large carp. Little Cache Lake can be found in Johnson County, northeast of Vienna. This shallow Shawnee National Forest body of water is 21 acres in size. This lake produces good largemouth bass and bluegill fishing, as well as fair fishing for redear sunfish and white crappie, and also offers channel catfish. In the Shawnee National Forest outside of Carbondale, in Jackson County, lies Little Cedar Lake. Feeding Cedar Lake, Little Cedar sits just above the larger reservoir, separated by a rocky waterfall. Little Cedar Lake is 84 surface acres in size and is best known for its largemouth bass and bluegill fishing.

The Little Wabash River begins in central Illinois and flows south through the Southern Region counties of Effingham, Clay, Richland, Wayne, Edwards, White, and Gallatin before spilling into the Wabash River near New Haven. Channel catfish are the most commonly caught species here, and fishing can be very fast from average-sized cats, but largemouth bass, bluegill, and crappie are popular as well. Jana Hirst lists the Little Wabash as a top stream for largemouth and spots, black and white crappie, channel cats and flatheads, and yellow and white bass. McLeansboro City Lake is located in Hamilton County, about a mile southwest of McLeansboro. This reservoir is 75 acres in size, with a maximum depth of about 23 feet. Best known for its excellent channel cat fishery, McLeansboro also offers bluegill, redear sunfish, crappie, and flathead catfish, and it also produces good largemouth bass fishing.

Mount Vernon Pond is located in the Mount Vernon Game Farm in Jefferson County, 3 miles south of Mount Vernon. This 2-acre body of water has a maximum depth of about 10 feet. The pond offers rainbow trout, largemouth bass, channel catfish, yellow bullhead, and bluegill fishing. Trout are stocked at Mount Vernon in spring and fall. Norris City Reservoir can be found in White County, just east of Norris City. This 22-foot-deep body of water is 28 acres in size with an average depth of 11 feet and about a mile and a half of shoreline. Norris Lake's channel cat fishing is rated as excellent. It has good fishing for large-mouth bass and bluegills, fair crappie angling, and also offers flathead cats and redear. In Gallatin County, 1 mile west of Omaha, Omaha Reservoir is 22 acres of water to a 16-foot depth, with an average depth of 7 feet and 1 mile of shoreline. The reservoir produces fair fishing for largemouth bass and black and white crappie but good fishing for channel catfish and bluegill.

Omaha Township City Reservoir is located 2 miles northwest of Omaha in Gallatin County. This 27-acre lake also has a maximum depth of 16 feet and an average depth of 7 feet. This lake produces excellent channel catfish angling, and bluegill fishing rates very good too. Good largemouth bass fishing is available, as well as fair fishing for white and black crappie. Pinckneyville City Lake is 3 miles northwest of Pinck-neyville in Perry County. This 165-acre body of water produces fishing that is rated as good for both largemouth bass and channel catfish, as well as an improving redear sunfish fishery. Bluegill and crappie are also available. In Gallatin County, southwest of Gibsonia, Pounds Hol-low Lake is another U.S. Forest Service body of water, coming in at 28 surface acres with a maximum depth of 24 feet and an average depth of 16. Pounds Hollow produces channel catfish opportunities rated as good, as well as fair largemouth bass fishing in addition to bluegill, redear sunfish, and black crappie.

Raccoon Lake is located east of Centralia, in Marion County, Illinois. A shallow yet fairly large lake at close to 1,000 surface acres in size, Raccoon has a maximum depth of 17 feet and a mean depth of 4 with over 16 miles of shoreline. This body of water produces good fishing for crappie, black and white, and largemouth bass fishing that rates good as well. Raccoon Lake also offers channel catfish, yellow bullhead catfish, bluegills, and large carp, and Boone LaHood rates it his top pick for white crappie. Two miles northwest of Ramsey in Fayette County lies Ramsey Lake. This body of water is 54 acres in size with a maximum depth of 22 feet and an average depth of 8 feet with about 3 miles of shoreline. Ramsey sports good bluegill, redear and channel cat fishing, as well as very good black crappie fishing. Largemouth bass, green sunfish, and yellow bullheads are also present. The lake offers fishing for saugeye as well, and it's the top pick of Boone LaHood for bluegill and redear.

Red Hills Lake is located at Red Hills State Park near Bridgeport in Lawrence County. This body of water is 41 acres in size with approximately 3 miles of shoreline. The lake contains largemouth bass, channel catfish, and black crappie. Bluegill and redear are also present, and it's Boone LaHood's top pick for warmouth. In Sahara Woods of Saline County, 5 miles west of Harrisburg, is Sahara Lake. The approximately 98-acre lake is surrounded by strip cuts, which altogether make up a total of about 270 fishable acres of water. Largemouth bass, channel catfish, redear sunfish, and bluegill angling all rate as very good on Sahara, and the crappie fishing rates good for black and white specs. Saline River feeds into the Smithland Pool of the Ohio River, along the border of Gallatin and Hardin Counties, at Saline Landing east of Cadiz. Big drum, gar, carp, and other rough fish are common. Jana Hirst lists it as a top pick for largemouth, black and white crappie, redear, flathead and channel catfish, and white bass, which all produce very good fishing

Sam Dale Lake is located 5 miles northwest of Johnsonville in Wayne County. This is a 194-acre body of water surrounded by forest, with camping available. Fishing is rated as very good for largemouth bass, channel catfish, and redear sunfish, and there is good fishing for bluegills. White crappie and pumpkinseeds are available, and the lake offers fair fishing for muskies too. Sam Parr Lake can be found 3 miles northeast of the town of Newton, in Jasper County. Sam Parr is a long winding lake, 180 surface acres in size with nearly 10 miles of shore-line. The lake has a maximum depth of 28 feet and a mean depth of 10. This body of water produces very good angling for largemouth bass and channel catfish, as well as fair fishing for white and black crappie. Bluegill, pumpkinseed, green and redear sunfish are available, as are carp. It's also Boone LaHood's top pick for hybrid crappie.

In Effingham County, 5 miles southwest of Effingham, lies Sara Lake, which has 614 acres of fishable water with an impressive depth for its size of 47 feet, averaging 19 feet. This body of water has 38 miles of shoreline. Crappie fishing is good here for both black and white specs, and the lake offers fair fishing for white bass and channel catfish. The largemouth bass fishery is improving. Bluegill, pumpkinseed and redear sunfish, and yellow bullheads can also be found swimming these waters, as can very large common carp. Sesser City Reservoir is a shallow 20-acre body of water, southeast of Sesser in Franklin County. This lake is surrounded by woods and sports very good fishing for largemouth bass and bluegills. Redear sunfish angling rates good here too, and black crappie and channel catfish are available as well. Sugar Creek Lake is located between Dixon Springs and Grantsburg in Pope County. This is a 96-acre body of water. Channel catfish and bluegill are available. In Hardin County, 3 miles northeast of Elizabethtown, anglers will find Tecumseh Lake. This body of water is 13 acres in size with a maximum depth of 16 feet and a shallow 4-foot average depth. Tecumseh produces good channel catfish angling and also offers large-mouth bass, bluegill, and black and white crappie.

Ten Mile Creek is located in Jefferson and Hamilton Counties, near Belle Rive. Auxier Creek and Rocky Branch flow nearby, and the State Fish and Wildlife area includes several stocked strip pit lakes for fishing. Largemouth bass and crappie fishing is good here, and channel catfish, bluegill, and other sunfish are available as well. Vandalia Lake is located just northwest of Vandalia in Fayette County. This 646-acre body of water has 12 miles of shoreline, and camping is available. Largemouth bass, channel catfish, and bluegill are regularly stocked at Vandalia, and Boone LaHood lists it as his top choice for channels. About 4 and 5 miles east of West Frankfort in Franklin County, anglers will find the pair of West Frankfort Lakes. New City Lake is 214 acres in size and Old City Lake has 147 acres of fishable water. New Lake provides very good fishing for largemouth bass, channel catfish, and bluegill, as well as fair crappie fishing, particularly for white crappie. In Old Lake, fishing for channel cats and bluegills is rated as very good, and it also produces good fishing for redear sunfish as well as for largemouth bass and black and white crappie too.

Southern Illinois is home to various small creeks not listed here, which can produce varied fishing opportunities. Fishing small unpressured streams can be especially productive if they are downstream of lakes, or even small farm ponds, where big fish sometimes escape during flooding. This region of the Prairie State is dotted with many small private lakes, ponds, and flooded strip pits. Many of these can and often do produce tremendous fishing for the owners, as well as their guests who obtain permission.

Anglers landed the state-record 2-pound, 12.3-ounce redear sunfish from the private Arcadia Lake in Marion Country Club, and the 3-pound, 8-ounce state-record bluegill that came from a Southern Illinois farm pond. For these and a full list of Illinois and North American fishing records and other great information, visit the state IDNR website at ifishillinois.org. Yes, world-class fishing for bluegills and several other sunfish species is available here, with many ponds, strip cuts, and small lakes producing high numbers of big specimens. These can be real sleepers. They often harbor giant bass, large crappie, big old catfish, and more in waters that are seldom targeted.

The waters of Southern Illinois are magnificent and their fisheries productive, each in its own way. Based on things such as size, depth, clarity, available structure and cover, and species, each also has its own challenges, as seasons and weather conditions change. And we'll dive into that next.

5

Season and Location

The season of the year and the stage of that season is the first thing to take into account when attempting to develop a hot fishing pattern anywhere, and Southern Illinois is no exception. Each and every water has a variety of specific seasonal quirks that are unique to that lake or river or reservoir. But there are many more that hold true to all waters and can be more or less counted on, at least enough to track the development of patterns for that particular time of year. Of course, it's not an exact science, no matter how much we wish it could be. Various fish behaviors can't be nailed down precisely to any particular set of dates or even exact water temperatures, as things change from year to year. But hiring a guide can really shrink the learning curve. This is especially true for developing seasonal patterns as our world continues to change radically.

Throughout my life, we've experienced the occasional year here or there that has produced significantly unseasonable weather. But this phenomenon is definitely increasing. The last decade of weather in this region, especially as it relates to fishing, has been downright bizarre. I can't count how many times my charter clients, guides, tournament partners, and other friends and family have been out on the water with me and brought up just how crazy the weather has gotten. Everyone notices it. But anglers, hunters, and all those who love the outdoors and spend a great deal of time out in it seem to notice these changes even more. We all talk about how you just can't count on the weather anymore. When March seems more like May and April feels like February, the fishing is going to suffer. Most years now September feels like

July, and we've donned the dive gear to swim with the fish numerous times in March and November without a wet suit!

Spring does seem more affected than the other seasons here in the bottom portion of the Prairie State. Spring fishing in Southern Illinois has been very consistent throughout most of my life. But it's been over a decade now since we have had two fairly consistent springs in a row here. We say that fall fishing is always good. In reality, there probably aren't any absolutes anymore. But it's almost always good in the lower land of Lincoln. Fall fishing seems less affected by unseasonable weather than the rest of the year here. But we still have to adjust more often than we used to. Just a few short years ago, Cade and I spent most of a day snorkeling a spring-fed stream in late October, without wet suits, and it was a lot more comfortable than being out of the water! Luckily for the fish, they are adaptive creatures. They will adjust to their environment as best as they can. Luckily for anglers, fish still have to eat no matter how crazy the weather becomes. But we have to pay close attention to everything in our environment and be willing to change the way we pursue these creatures for success.

Studies have shown that some fish are homebodies and generally stay in the same particular areas throughout the year. But we also know that other fish tend to roam. A much larger mass of fish will choose to migrate from place to place with changing seasons, setting up shop in optimal locations. Movements depend primarily on water temperatures, length of daylight, and prey migration. So, while there are no hard and fast rules on when to go where on the water and when to do what to catch fish, we can follow the general guidelines of seasonal fishing based on the time of year. Then we adjust whenever Mother Nature tosses us a curveball.

Early and Middle Winter

In early winter, as Southern Illinois water temperatures drop significantly and fall into the upper 40s, we see major changes in fish behavior. This continues on through the majority of the winter period here as water temperatures plummet into the 30s and down to the freezing mark. From a location standpoint, a significant migration takes place. Most baitfish species and the majority of game fish move from shallower and more confined locations in creeks and river arms, coves, pockets, and bay systems into deeper and more open main basin type areas on lakes and reservoirs. In rivers and large streams, fish will move out of feeder creeks, backwaters, side channels, and connected chutes and oxbows into the main river channel itself, or into deeper holes in the lower sections of the feeder streams.

In lakes, many fish will spend much of the wintertime suspending in moderate to deep water. Of course, this can occur on almost any body of water, but the deeper and clearer lakes like Grassy, Kinkaid, and Kitchen tend to produce better action from suspending fish than shallower and more stained waters like Crab or Rend, especially with regard to black and temperate bass. Suspending fish may relate loosely to major structural elements such as long main lake points, distinct break lines (drop-offs), humps, holes, and creek or river channels. But some may not necessarily relate to any structures at all and instead cruise flat featureless basin areas halfway down in the water column or more. And this depends largely on pelagic baitfish such as shad, herring, and shiners. When large schools of baitfish are concentrated in the area, gamefish will follow them wherever they swim. When baitfish are scattered and large schools are difficult to locate, finding these prime structures and fishing closer to them becomes more important. In these cases, fish are much more likely to hold right on the bottom or along breaks and points and channel edges, even hiding in and around cover. In rivers, fish primarily tend to simply hold in deeper holes or around current breaks along channel edges to conserve energy. It is during this time period that many of the largest blue catfish are caught in deep holes of the Ohio and Mississippi Rivers. Most of these fish remain out of the main flow and allow prey to wash downstream to them.

While aquatic vegetation is important to fishing success at other times of the year on many Southern Illinois lakes, the opposite is true in winter. Healthy, green, growing vegetation pumps fresh oxygen into the water through photosynthesis. Brown, dead, and dying vegetation, however, has the opposite effect, removing dissolved oxygen from the water through the process of decay. In a matter of just weeks, it's common to see fish go from heavily utilizing vegetation to avoiding it altogether. When targeting game fish that are not suspending in open water, you will find that the best wintertime structures have both rock and wood on them. Wood cover such as standing timber, stumps, lay-down trees, logs and brush, or even manmade cover designed to mimic wood, like plastic trees and brush piles, are very attractive to most game fish species throughout winter. Rock can make an area like this even better.

Rocks can also be an excellent cover option that Illinois gamefish utilize to hide from their prey, as well as to avoid predators themselves, especially when they are young and smaller in size. Rocks also warm quickly on sunny days and tend to retain heat well. Locating warmer water in winter can mean finding much more aggressive and active fish and therefore hotter fishing when it's cold. Natural chunk rock, ledges and boulders, as well as concrete pilings and human-placed riprap are used heavily by gamefish in winter. On warmer sunny days in this region, fish will often move up out of the depths to hold and feed in shallower water. This is especially true when these rocky spots are close to a sharp drop-off to deeper and more open water. Here, fish can quickly and easily slide back down as air and water temperature drop at night or cool during a weather change.

Fish activity slows considerably in wintertime throughout Illinois, much more so than in waters throughout the Southern United States. Luckily though, the decrease is less pronounced here than it is in the central and northern parts of the state. But with cold water temperatures comes slower fishing action. The big advantage, however, comes in a massive reduction of boat traffic and fishing pressure. Having fewer anglers on the water is always a benefit, and fish become less conditioned to being targeted. While much fewer fish are typically caught, they can be more impressive in size, and some true giants have been landed from frigid water. Water in the low to mid-40s is much better than water in the 30s. But even when the temperatures dip down around the freezing mark, big fish can still be susceptible. This is especially true with some sunshine and the right wind.

Still, reduced activity often means slowing way down in presentation speed here. From a casting perspective, slow cranking, slow rolling, crawling, dragging, and even dead-sticking presentations with artificial lures become more effective at this time of year. Still fishing, or slow drifting with live or dead natural baits depending on the species, is also an effective tactic. Slow trolling can produce well too, if speeds are kept low enough for the particular fish targeted. Southern Illinois crappies and walleyes that are trolled up in winter usually require an extremely slow trolling speed, under two miles per hour—and sometimes below one mile an hour. But muskies, stripers, and even bass can be taken at higher speeds here.

Active wintertime bass and stripers sometimes take lures trolled at 2 to 2.5 miles per hour, while muskies will sometimes chase down big baits running at 3 to 4 miles per hour or even faster. Winter is also a great time for vertical jigging. Because winter fish tend to concentrate in deep water, holding the boat right over top of them while dropping a lure to slowly jig up and down in their faces will often produce the hottest bites when it's cold. Always remember that less is usually more when it comes to presentation cadence and speed in the coldest parts of winter.

Late Winter and Early Spring

As the middle portion of winter passes by and late winter arrives in Southern Illinois, things can change in a hurry with warming waters. As the days lengthen and sunlight has more time to warm the shallows, many fish will begin migrating again—as long as the region doesn't experience unseasonably cold weather, that is. After lakes and rivers have been down in the 30s for a time, a significant fish migration can begin as soon as the waters climb back into the low 40s. This of course varies a bit from water to water and species to species. But this is an exciting time of year. As winter begins to release its icy grip and we look forward to lots of pleasant weather coming very soon, the greatest thrill is the chance at the biggest fish of the year.

We do catch big fish in Southern Illinois all year round, but prespawn beasts are heavier than they will be at any time of year. Both males and females have been bulking up. After feeding heavily in the fall and expending very little energy through the coldest period, they're heavy and healthy. As activity increases, they put the feed bag back on again to pack on a little more weight before going through the rigors of spawning. Speaking of spawning, adult females are even bigger, huge in fact, full of eggs and maxed out. Catches during this time frame make for awesome photos, and replica mounts for the wall. Its SI trophy time!

Even within each particular species, some fish will begin to migrate before others. And some areas of a particular water will see more migrating fish than other areas. But as a general rule of thumb, when the water temperature climbs back up in the 40s, fish begin moving. This may seem strange to some, since these temperatures may have triggered the opposite effect in early winter. But length of daylight plays a role too. Fish sense the changing of the seasons in multiple ways and change behavior accordingly, for feeding and for reproduction. Cold-water species like muskies and walleyes move earlier, often as soon as it hits forty degrees, while many warm-water fish like largemouth bass and catfish might not be really moving until the waters hit the upper 40s. As waters climb into the low 50s, the bulk of fish finish migrating and arrive in the shallows where they will spend a considerable amount of time.

Late winter and early spring mean an increase in fish activity again. Fish become more willing to strike faster, horizontal moving presentations. This is good, because they don't bunch up nearly as much at this time, and the vertical jigging bite falls off with few fish suspending in open water. Vertical tactics still produce fish, as they do throughout the entire year. But it's vertical fishing with a cast-and-retrieve presentation rather than getting over top of schools of fish and repeatedly jigging lures up and down in one spot. When fishing vertical with a casting approach, lures like jigs are cast out and allowed to fall to the bottom, where they are dragged and hopped, on and off structure, as they are retrieved back to the boat. The main difference is that we likely won't spend much time with strict vertical jigging presentations over the top of fish in open water.

Slower retrieve speeds are still the norm in late winter in this part of the state. But this slowly changes in the transition to early spring, when more standard moderate retrieve speeds often work best. At this time, we're usually fishing shallow to mid-depth zones rather than deep water. But the areas we're searching are still not far from fairly sharp drops to deeper water, where fish expend little energy to slide into the depths with passing cold fronts.

On most Southern Illinois waters, the bulk of the fish travel along predictable migration routes. Look for breaks from shallow to moderate depths, where the contour lines bunch up closely on a map. The first major drop, out at the edges of shallow flats and big shelves where it falls fairly quickly into moderate depths, often holds these migrating fish. As they come in off the more open waters of the deep main lake or the main river channel, fish will use these breaks like highways, often stopping to hold on changes along the way. Irregular sections of the break will often harbor numbers of fish. But anywhere that you find a point, inside turn, or bend in the break line itself is a likely spot for fish to stop and hold during their travels.

Middle Spring

So, where are SI fish headed? Back into the same kinds of locations they left as early winter came around. Throughout the majority of the spring period, or when the water temperatures range from the low 50s to the middle 70s, the majority of Southern Illinois gamefish on lakes spend their time up in river arms, creeks, coves, pockets, and bays. In rivers, they'll be in feeder creeks, side channels, backwaters, connected oxbows and chutes, any shallow flooded areas, or in normal levels, shallower holes, shelves, and flats outside the main river channel. Springtime gamefish search out the protected locations where they will spawn and carry on the cycle of life. The middle spring offers a shot at fat prespawn giants of most species. Any of these confined shallower waters can hold plenty of active fish throughout spring and produce feverish action at times. But from late winter through the middle portion of spring, it's always best to head to the north side here.

The north or northwest portions of lakes and rivers receive more sunlight than do the waters in other areas, and therefore warm more quickly. The water temperatures can be significantly warmer on the north and northwest sides of fishing areas. They often harbor more fish, and fish that are more aggressive. Still, look also for smaller spots that are protected from strong winds that might blow in colder main lake water, as well as areas out of the heavier current flows on rivers. Confined locations, and those with either dark bottoms or rocky shorelines, will often warm faster and retain heat better. Rock, wood, docks, and other man-made cover objects continue to hold fish at this time. But weeds, in waters where they are available, start to become more important to fish with warming waters and new growth.

Sometimes, getting far up into creeks and backwater areas of SI lakes and rivers can lead to incredible fishing. This can happen at any time of year, anywhere across the continent. But in Southern Illinois, it's most productive in spring and fall when fish move shallow and into more confined locations. It's even more practical in spring than fall though, simply because spring rains usually raise water levels, making more shallow areas accessible even if few take advantage of it.

Most SI anglers bypass such places because they take time to get to. But, when fishing pressure is high, getting away from the crowds can be a ticket to success. We've caught countless big fish of many species by going as far up creeks as we can, or by easing our way through tight areas to get into marshy swamps or hidden bays and coves across Southern Illinois. Shore-bound anglers can often hike into remote areas of larger waters. Paddling gets us into these goldmines as well, and I thoroughly enjoy silently fishing from my kayak or a canoe. I've also done some fishing from friends' jet boats, which can run fast through incredibly shallow waters.

While I prefer to use my 21-foot bass boat and 250-horsepower outboard most of the time, there have been many instances over the years when I've chosen my backup bass boat instead. At about 17 feet, with a 150-horsepower engine, this boat gets into slightly skinnier water than the big one. And more importantly, it maneuvers around tight corners in narrow creeks a little easier. The walleye boat is even smaller. Still, even large boats can access places where the vast majority of the angling crowd never ventures. And I've slowly inched my way far up creeks and backwaters in the big one. I'll usually go as far as I can with the outboard, using the trim and the jack plate to get it up high and out of danger of submerged obstructions. Then, the trolling motor takes us even further up to fish that may not have seen a lure for quite some time.

As the waters warm, anglers can progressively fish faster and faster. It's a great advantage when trying to locate numbers of active fish and when trying to search out and select areas that are holding larger-than-average fish. With the exception of catfish, most gamefish species go for artificial lures the majority of the time now. While live baits can certainly still take plenty of crappies, bass, and other species, using artificial lures will always be a much quicker method of fishing. Lures allow the angler to cover more water more effectively and get in front of more fish, ruling out less productive areas and narrowing the search as efficiently as possible. Moderate- to high-speed presentation is usually best. This is often the case. But exceptions do exist, such as when a cold front moves through the region or sometimes on busy weekends when fishing pressure is very high. Then again, maybe that means it's time to head up a backwater!

Late Spring and Early Summer

As the spawn finishes up for most Illinois species here and we find ourselves in that magical late spring to early summer transition period, fish activity continues to pick up with most species in water temperatures ranging from around the middle 70s to middle 80s. Another migration occurs at this time, as most fish will begin to move out of the confined and protected areas and back out toward the wide main river, or the deeper and more wide-open basin areas of lakes. Most fish do not move into deep water yet, just closer to it. Fish can often be found spread out and feeding across large shallow flats and shelves, along gradually tapering shoreline banks and points, and on shallow humps. Some fish also suspend over deep water at this time here, but remain within the top 10 or so feet of the water column. One would think that the fish moving into such large and open areas would make fishing more difficult. But the significant increase in activity seems to offset that. With so many fish active and aggressively hunting for so much of the time, some of the hottest action of the year occurs.

Southern Illinois fish that lost weight during the spawn are hungry. They're in a hurry to put that weight back on, and more. The well-discussed period of sluggishness that fish go through immediately following the rigors of spawning is relatively short-lived. And it's quickly followed by frenzied feeding activity. All forms of cover are still utilized by most fish that aren't suspending. And locations with two or more different types of cover are always highly attractive. But weeds and grasses really become important to most species now in waters where they are available. Thick, green, growing aquatic vegetation, pumping out lots of oxygen, holds massive amounts of the prey that gamefish feed on. And this cover allows these predators to hide and hunt very effectively. For the suspending fish in open water, late spring and early summer bites are often red hot. Faster presentations become even more effective, with horizontally moving lures producing far better than vertical ones most days. This is also one of the two hottest times of the year for a topwater bite, as many species of fish feed heavily on and near the surface.

Middle Summer

The middle summer can produce some excellent fishing in our region. It's certainly not as great as late spring and early summer, when water temperatures are within or close to the optimal temperature ranges of most gamefish. But it is still exciting. Water temperatures typically range from the mid-80s to the low 90s but can sometimes climb into the mid-90s during longer periods of hot and dry weather. While waters can sometimes get a little too hot for things to go perfectly, the high temperatures do mean high metabolisms in fish and a need to feed. At this time, some fish can still be found in the shallows, especially on and near the main lake. But typically, these fish will move even closer to a sharp drop-off to deeper water.

Instead of being scattered across huge shallow flats, with many positioning far from the depths, the shallow fish will likely hold at the edges of the flats near a sharp drop-off or along the break line of a steep sloping shoreline bank, still shallow but close to deep water. Others move out further on points. Some may hold on the tips of shallower points while others may hold along the spine, but on the sharper sloping side. Rocks and wood still hold fish in the shallows, and wood is especially important on waters with little vegetation. Even without any other cover though, weed beds and weed lines or patches of grass will hold plenty of fish all by themselves. Summertime fish just love the salad!

While some fish remain fairly shallow, many also choose to spend more time in deep water. Excellent spots to investigate include the tips of deeper points, deep humps, old roadbeds, mid-depth flats, and the break lines to deep water at their edges, as well as creek or river channel edges. Bends and intersections of creek and river channels are particularly attractive to fish. This is great time of year to fish islands too. Islands hold lots of summertime fish, as long as they have at least one side that falls fairly quickly into deep water. Where available, deep-water fish will use deep weeds heavily. The deep weed edge, where the weeds stop growing as the bottom falls off, is a classic middle summer location. Various forms of rock and wood cover also hold fish, especially rock ledges and standing timber in deeper water. Concrete bridge pilings, the deep ends of boat ramps, and similar cover can hold fish with deeper water nearby. Docks over deep water can also be magnets for every species of gamefish now. The best docks are over the depths but not far from shallower water.

Just like winter, summer is also a fantastic time to search open water for suspended fish. Almost all species of gamefish will suspend through the middle of summer at some point. And while there might be plenty of fish in shallow cover or on deep structure, the open water bite can be even hotter. When gamefish suspend and chase shad or other schools of pelagic baitfish, the action can be incredible. One minute you can be yanking up fish after fish while vertically jigging deep water and the next be casting to schools of gamefish busting shad on the surface that they chase up out of the depths. It can be heart-stopping!

Still, whether you're casting, trolling, or jigging artificial lures, dragging or drifting live baits, or using a combination of tactics, a medium to high-speed presentation is usually the ticket on Southern Illinois waters now. We seldom have to slow down and finesse fish unless a massive cold front comes through. Speed of course is relative. While keeping live baits moving for catfish is still a lot slower than casting spinnerbaits for bass, it's still best to start out by fishing as quickly as possible and only slowing down when necessary. It's amazing how fast you can drift and troll baitfish for cats. But even when you're still fishing specific spots, it's usually better to spend less time per spot when it's hot. Quickly move to the next, until numbers of fish are contacted.

Summer fishing is a gas, with one exception: muskies. Timing varies, depending on the weather and water temperatures, from year to year. But during most years, we stop muskie fishing in Southern Illinois by about the middle of June, and we start back up in early September. While there is no official closed season here, we give the muskies a break during the heat of summer. None of the muskie guides will run charter trips for this species once water temperatures climb out of the low 80s, and we encourage everyone else to follow suit. Being northern cold-water fish, muskies stress far more than most species in summer. An occasional fish can die just from the heat alone, but most adult skis are susceptible to premature death when caught in the hot water.

We're all very careful with these fish throughout the year here. We use huge nets, keeping them in the water during the unhooking process. We pull them out only for a few quick photos after getting everything ready beforehand, and then release them quickly. But even with the best efforts, in July and August many of the SI muskies caught here will die. Some will not be able to recover and swim off on their own, going belly up as anglers attempt release. Others might barely swim off but will die later and float back to the surface. Typically, the bigger the fish and the longer the fight, the harder it is for them to survive being caught. Luckily, muskie fishing would be difficult in summer anyway. Most of the adult fish become less active, lethargic even, and don't feed often. They spend much of the summer attempting to regulate their body temperature while also trying to get enough oxygen.

If we do occasionally hook a muskie while targeting other species, we usually give it a little slack line and try to let it shake off. If this doesn't work, we fight the fish in as fast as we can. We then quickly unhook and release it boatside without ever removing it from the water. One angler can photograph the animal while the others unhook and release it. Time is of the essence. And for fish that are difficult to unhook, it can be better to simply cut the hooks off in order to speed release. This is one of the reasons why most muskie anglers keep a good pair of one-handed bolt cutters on board. It's a great idea for any angler targeting any species though. As those of us who have ended up with a hook deeply embedded somewhere in our body and an angry fish pinned to the other end of the lure can attest, even if bolt cutters cost $1,000, it would be well worth the investment!

I've discussed summertime southern muskie mortality while giving seminars and on television and radio programs. And I mentioned the topic in some newspaper columns I wrote. Later, I talked with fisheries biologists about this from multiple states and got into greater detail about the subject in articles I wrote for *Adventure Sports Outdoors*, *Esox Angler*, *MidWest Outdoors*, and *Muskie* magazines a number of years ago, as this is not just a problem in Illinois. Most muskie lakes and reservoirs across all states in the southern muskie range experience this issue. Luckily, we've been able to teach much of the angling community about these dangers. As a result, most southern anglers do not target muskies during the hottest portions of the year. And the majority of northern anglers make their trips south for muskies from fall through spring instead. It's made a difference in the health of our muskie fisheries. Luckily, fishing is often good for most other species during summer.

Late Summer and Fall

Now onto my favorite time of year in this wonderful place. On my travels, I have found different times of year to be better for fishing in various places. Summer is my favorite season in Canada, while winter is when I prefer to be in Mexico. More specifically, I prefer early summer fishing in Alaska but early spring in Costa Rica. September is my favorite month on Missouri's Ozark streams, while April would be my pick for lakes in eastern Kentucky. As much fun as it can be to fish Southern Illinois waters in summer, and as amazing as the fishing can be here throughout spring, when the weather cooperates, my favorite time of year in the bottom portion of the Prairie State is the long period from late summer to late fall. When waters fall into the low 80s, all the way down to about 50 degrees.

During this wondrous time period, fish activity increases here, and then increases, and then increases some more. There are a great many reasons why this time frame is so good. For one, the vast majority of Southern Illinois gamefish become fairly predictable, following the same cycles from year to year. They become extremely aggressive, feeding heavily and feeding often as they attempt to bulk up and make their greatest gains of the year. We see a major reduction in overall SI water use at this time as well. Pleasure boating traffic decreases significantly, and so does the fishing pressure. As family vacations wrap up and kids go back to school, all activity on lakes and rivers decreases and the waters get quiet and more peaceful again.

There are far fewer fishing tournaments held here during the late summer and fall than there are in the springtime and early summer. Additionally, numerous fall hunting seasons in Illinois and surrounding states, for everything from squirrels, rabbits, and doves to deer, turkey, and waterfowl, cause a lot of anglers to hang up the fishing rods for the year. This all boils down to fewer boats and less pressure on our fish, which means better fishing for the rest of us. I am a hunter and I do love the sport. But since I was a kid, hunting always took a back seat to fishing.

The weather is drier and more stable than in spring, especially these days, and this is clearly evidenced by lake and river levels. All dedicated anglers know from experience that we typically deal with higher water levels in spring and early summer but lower water levels in late summer and throughout fall. Additionally, with increased activity and a natural cooling process, SI gamefish become less affected by cold fronts and poor weather conditions. Unlike late winter and spring, fish are not at all influenced by the spawning process in late summer and autumn, as they're focused solely on feeding. More fish begin chasing shad, herring, sunfish, chubs, shiners, and similar baitfish species as the primary food source. So more different species of fish are typically caught during an outing. And water clarity typically increases here, while at the same time fish show a preference for more visual lures and presentations, thereby producing some of the most heart-pounding, excitement-filled fishing of the year!

As late summer rolls around, we again see another significant migration of the vast majority of gamefish species in our area. Most of these fish leave the deeper and more open main river channels and main lake-basin-type areas and move up into the shallows again. Without unusually high fall rains, oxbow lakes are normally cut off from main rivers, and flooded woods and fields are dry. But feeder creeks, side channels, backwaters, shallower holes, and sandbars out of the main river channel become focal points in larger flowing waters. In lakes, creeks, and river arms, bays, coves, and pockets hold most of the fish again. Many of these fish will still utilize cover. Weeds and grasses are used extensively until the later portion of this time frame, when growth stops or slows significantly. Fall fish love weeds on the Southern Illinois waters that contain them. But all forms of rocks, wood, and man-made cover objects are utilized here as well.

With high fish activity during this entire time frame, moderate to high-speed presentations are again the norm from day to day, with a couple of exceptions. Crappies will still chase down fast-moving horizontal lures. But many still hole up in cover, where vertical still fishing is as productive. And while still fishing for cats is best most of the time, drifting and dragging live and dead baits does produce big fall whiskerfish. We also seem to catch more catfish on artificial lures than during other periods, lending even more to the multispecies advantages of the late summer and fall fishing season. However, all of the black and temperate basses, walleyes and sauger, muskies, trout, and sunfish all chase down horizontally moving lures with regularity, and most respond best to retrieve speeds in the moderate to fast range.

That's a Wrap

Finally, as the year comes to a close here, waters get cold. Fish activity slows back down, and the late fall gives way to the early winter period again. SI fish migrate back out to those deep, open main lake and main river haunts, where they will stay until things start to warm again and the whole cycle of life on the water repeats itself. Oh, how I love the seasons. I will admit that I greatly enjoy cool, warm, and even hot weather, while not being much of a fan of the cold anymore. But I do love how things change here.

I love to watch the new green burst loose every spring, and how the brilliant oranges, yellows, and reds turn the forest to fire in autumn.

I long for the feeling of warmth after that first reprieve in late winter, and how those cooling breezes of fall cause you to just stop everything you're doing, close your eyes, and breathe deep. I yearn to jump into the refreshing Southern Illinois waters in summer, to swim and dive with the fish. And, I look forward to seeing the way our waters look with just a bit of ice, when all the leaves are gone, eagles seem to soar everywhere, otters once again play out in the open, and a dusting of new snow opens up the woods surrounding lakes and rivers, causing deer, coyotes and bobcats to stand out against the white background. I love it all. Take the time, each and every year, to become intimate with each and every season. Take Southern Illinois and its magnificent seasons into your heart, and grow your connection to the wild.

6

Fishing Conditions

It is said that 10 percent of anglers catch 90 percent of the fish. This common saying absolutely rings true around the world, and Southern Illinois is no exception. Regular time on the water, years of fishing experience, the acquisition of knowledge, and the refinement of skill all play critical factors in an angler's ability to make it into that 10 percent club. But there is absolutely no question that connecting with nature helps an angler judge fishing conditions and develop patterns. Time in the outdoors can be a spiritual experience. It can touch you, deep within your own soul, clearing away the mess that the rest of the world leaves, sharpening your mind and strengthening your heart. It can make you new again, whole and complete. Immersion into nature can make you one with the wild and its creatures, if you allow it to. And that's the real point: if you allow it to.

For real anglers, true lovers of the great outdoors, those of us who are sincerely dedicated to the perfection of our sport, this art, this age-old method of survival passed down from our ancestors, we feel so comfortable on the waters, in the wilderness, with wild creatures, so at home that the civilized world will never provide us with what we truly need in this life. For us, nature is the real world. And those of us who live to immerse ourselves in it develop and constantly hone our senses, senses that other people will sadly never know or understand. For those of us who are truly passionate about fishing and other sports and traditions of the outdoors, we feel it, we breathe it, we live it.

Luckily, anyone can work to develop a stronger connection to the natural world. The best way of course is to spend more time in nature.

But it might be just as important to also change the way you spend time in nature, letting go of stress, clearing the mind of everyday garbage, not looking at the phone. With long-time clients, it's been exciting to see firsthand the development of their connection with nature, the refinement of their senses, and the development of solid fishing skills, all of which have helped them make it into that 10 percent club.

It happens so frequently that I cannot recall how many times it's occurred, but I will often notice a minute change in the conditions of the natural world that no one else in my fishing party or hunting group picks up on, like decreasing humidity or a minor increase in wind speed. Nor can I recall just how many times such a realization has caused me to make an adjustment, like changing tactics or moving to a different location, putting us into action with fish and game.

While guiding a charter trip on Crab for bass one day in tough conditions, I had a group that had only caught a dozen or so average-sized SI largemouth and a couple of bonus crappies that were nothing to write home about. I noticed that the wind direction seemed to be changing just a bit, so I fired up the outboard. We moved to an area where the shorelines would be getting a bit more chop, around one side of the secondary points. The move paid off big time. We easily started catching triple the number of bass as we had been, along with a number of oversized green sunfish that happily attacked full-sized largemouth lures. And we got into a school of nice white bass too. We ended up landing several big largemouth. And one of my clients caught her biggest bass to date, a 7-pound, 2-ounce trophy that she released and had a replica mount made of for the wall. All of this from one move, based on a tiny change in the conditions.

Consider all of the fake and useless things of our modern world, the distractions that do nothing for us and the false connections that only distance us more from what matters in life. Value what truly matters and make real connections instead—to the wild and to wild creatures, to fellow outdoor enthusiasts who harbor the same passions as we do, to family and friends we share our love for the outdoors with. There really are no words to put the depth of such things into proper perspective. I live for times when I turn off the computer, walk away from the television, silence the smartphone and just be in the wild, be where I was made to be.

Getting out onto the water, and into the wilderness, is medicine like no doctor on earth can prescribe. When you get out there, be there. Stop everything and just look around. Be quiet and be still. Breathe deep, the clean country air. Breathe it all the way into your soul. Shut off the ridiculous world of overwork and overstress, of a crazy-fast pace and a multitude of mindless distractions and responsibilities. Turn off the radio, put the phone away, unplug. Allow yourself to decompress, and then adjust to your natural surroundings. You'll likely feel better than you have in a while. And you'll likely start to notice all kinds of things in this real world that you wouldn't have otherwise. Then, drink it in. Drink in real connection.

When the mind is clear and the connection to the wild has been made, what all anglers should be looking for anytime they hit the water is called *the pattern*. The pattern consists of a vast collection of information that is required for consistent success in our sport. Most specifically, it is how fish will behave on a given day based on instinct. If we know why these animals do what they do to survive, we can determine how to go about finding them and getting them to strike our offering. A haphazard attempt is rarely productive. Stopping the boat at the first point one comes to and randomly picking any old lure from a tackle box is a recipe for failure. Instead, we go about this scientifically and strategically. There is a method to the madness, a reason for everything successful anglers do. And no detail is insignificant.

Sure, everyone gets lucky now and then. Anglers who rarely fish sometimes happen onto red-hot bites, where they catch high numbers all day long. And first-time anglers have landed once-in-a-lifetime trophies here and there throughout fishing history. But these rare instances are the exception, not the rule. Luck plays a tiny part in this age-old game of man versus beast. Knowledgeable anglers who regularly practice their craft and always work to develop patterns instead of hoping for luck are the ones that score over and over again.

Observing and utilizing all of the information at hand on a given day is how to figure out a pattern. It's how a pattern is made, or built, so to speak. Anglers must first take into consideration all of the current conditions. Things like season, water temperature, water clarity, stratification, wind, fronts, water use, and fishing pressure all play critical roles in fish behavior. This affects, first, the most important factor for starting to develop a pattern: location. Before anything else, we must find the fish. But not just find them. The key word here is not *fish* but *the*, as in the right fish.

Locating even large concentrations of inactive fish that are unlikely to bite does little for us. Finding fish in a neutral mode of activity is certainly better than finding negative fish. But what we are after are active fish, fish in a positive feeding mode, fish that are aggressive and hunting. That, my friends, is what can lead to an incredible day on the water here in Southern Illinois! The hottest part of summer and, even more so, the coldest portions of winter keep fish in a negative mode more often. Negative fish usually hold in deep water. Deep of course varies. Winter fish might hold in 40, 50, even 70 feet of water or more on Devil's Kitchen, Kinkaid, or the Mississippi, while deep-holding fish on shallow waters like the Cache, Centralia, or Forbes will likely be somewhere in about the 8-to-20-foot range. Waters with more moderate maximum depth zones like East Fork, Egypt, Newton, and sections of the Big Muddy will see lots of fish move into the 25-to-40-foot range or so, depending on water levels. These fish are more likely to set up somewhere rather than staying on the move. And they don't feed much.

Active fish, on the other hand, often hold higher in the water column on any lake or river in our region. They're aggressive. They move around more and feed often. In the spring and fall, we find larger percentages of active fish more often, since water temperatures are closer to their preferred range. The more ideal a fish's environment is for a particular time of year, the more often they will be actively feeding. So first, we take all of the conditions into account and start searching. Then we begin to employ various methods, techniques, tackle and lures and baits to tempt the fish we've found to bite. If they are active enough.

Once we know where the concentrations of active fish are, location becomes apparent. The lake or river areas where most fish are hunting, such as main lake basins and river channels or creek arms, side channels, coves, and bays, the water depths preferred; the types of structure they're holding on (measurable changes in the bottom of a water), such as points, flats, and drop-offs; as well as the kinds of cover they're using, like boulders, standing timber, or weed beds, all make up the locational element of a pattern.

Once on the spots, we must determine what the fish will most readily bite on. Trial and error must be employed. But this process can be significantly shortened by factoring in the conditions. Will the fish prefer horizontally moving lures or vertical baits, a loud rattling presentation or a silent and subtle one? Will it be a fast retrieve speed or a slow one today? Are more fish attacking topwater lures or deep-diving crankbaits, jigs, or spinnerbaits, jerkbaits or spoons, lizards or swimbaits, dough baits or night crawlers, suckers or crayfish, lively shiners or cut shad? Larger or smaller sizes, with scents or without, with a steady cadence or an erratic one, power or finesse tactics? Is light tackle required or can we use heavy gear; will braided lines work or is fluorocarbon needed, casting equipment or spinning gear? What color patterns are best today? We'll delve deeply into the various lures, tackle, and equipment choices in chapter 8.

Of course, veteran anglers have a clear understanding of all the "shop talk" here. For the new anglers out there, it might sound a bit intimidating. But not to worry, there's a quick solution. With the sheer variety of products now available, those brand-new to the sport will want an effective crash course in lures and tackle firsthand. First, order some free fishing catalogs from Bass Pro Shops and other retailers. The old classic products that have produced for decades can be found in these, as well as the hottest new stuff that is raging on the tournament trails. One can familiarize themselves pretty well through this free, one-of-a-kind resource. These have more information and greater detail on product than you'll find anywhere.

Step number two is to go into a tackle shop or sporting goods store. Take some time to browse what's available, and then ask someone working at the store for some help. The folks I've talked with in these kinds of stores are some of the most friendly and helpful you'll come across in retail. Most will talk your ear off and provide a lot of great information about lures, tackle, and equipment while you've got it right there in your hand. It's a critical part of the learning process for brand-new anglers. Cost is a factor for all of us. Brand-new anglers will probably want to start off fishing for panfish and then catfish before moving on. Muskie fishing and bass fishing are the most expensive fishing sports to get into in freshwater, but effective gear for most other species is fairly inexpensive.

But back to it. At the end of our search when we've located the active fish, we then work to determine exactly what will make them bite. Once we determine the best way to get them to bite, then we have our pattern. Every point mentioned above must be determined in order to nail down a precise pattern and experience great success on the water. All of these are dependent upon the conditions. And all of this comes far more naturally to an angler that unplugs from the rest of the world and instead plugs into the natural world. Playing on the instincts of the fish, figuring them out, and coaxing them to come out and play with us is the name of the sportfishing game. It all starts with determining what the conditions are that we're given each day on the water and reacting to them. And determining and reacting to those conditions starts with developing one's connection to the natural world.

Water Temperature and Clarity

Most fish species are ectothermic poikilotherms. In simpler terms, they're cold-blooded. But there's more to it than that. Cold-blooded doesn't mean that they remain cold all the time. They adapt to their environment. Most fish in the heat of summertime will be very warm, hot even. Their bodies are usually at or slightly above the temperature of the surrounding water. Most fish have darker-colored backs and lighter-colored bellies. This is an advantage to both predator and prey, since they become harder to see from above and below. But there is another advantage, especially in the cold months of the year. A darker-colored back also absorbs heat from the sun rather than reflecting it. During daytime hours and especially on bright sunny days, many fish will move into the shallows to work on their tans. They use the sun to warm up, which boosts their metabolism and activity level.

Fish feel temperature changes far more than humans do. So, while humans wouldn't likely notice a minor change in temperature of just a few degrees, fish will feel a minor change considerably. Even just a difference of one or two degrees will often cause many fish to move into that warmer or cooler water, as well as to become more active and more susceptible to bite. Water clarity affects the way fish behave as well. And anglers must adjust their methods to compensate for changes in clarity, especially with the extremes. Most species can see a significant distance in ultraclear waters. In heavily stained waters, however, the suspended sediment in the water significantly limits vision. In this situation, fish rely much more on their other senses. They rely heavily upon their hearing and their ability to detect vibrations in the water through their lateral line. Basically, they hear the sounds and feel the movements made by their prey, and other creatures in the environment. While some regions of the continent have waters that are mostly clear, Southern Illinois has more stained waters than clear ones. But even some of the more heavily stained waters will clear up after periods with little to no rainfall. Kinkaid gets very clear at times, especially in the lower third from the main lake to the dam. The same is true of Cedar, some of the Pyramid Lakes, and other still waters across Southern Illinois. Some small farm ponds in the region can get really clear from late summer throughout the fall and well into winter. Even the often-muddy Mississippi River can be surprisingly clear when it's low, especially in fall. Boating or wade fishing the sandbar flats of the Mississippi where creeks flow in will show this easily. In most cases you can see several feet down to the bottom in the main river, but you can't see a couple of inches below the surface where the creek comes in.

From an artificial lure presentation standpoint, warmer waters and dirtier waters require the use of larger and noisier lures. While cooler and clearer waters tend to produce better with smaller and more subtle options. For instance, crankbait choices for largemouth bass and walleyes vary greatly. In warm and heavily stained water, the best choice is nearly always a bulkier crank with a wide wobbling action. This creates a lot of vibration that fish can feel with their lateral lines. They can more easily home in on this type of lure when they're active but can't see well. In cold clear water, though, a thinner crankbait with a tight wiggling action is more natural in appearance and creates less water movement. It works great on less active fish that can use eyesight exclusively to hunt and attack prey. The same works with spinnerbaits for muskies and stripers. Larger lures with round Colorado-style blades displace more water and produce more vibration, making them ideal in dirty water. A smaller spinnerbait with the thinner willow-leaf-style blades produces more flash and less vibration, ideal for clearer waters.

You can also apply this when selecting soft plastics that are ideal for, say, smallmouth and spotted bass. A thin 4-inch straight tail finesse worm that works great in cool clear waters might not produce at all in warm muddy conditions. But a thick-bodied 5-inch beaver might get hammered in hot dirty water but fail to draw a strike where it's cold and clear. This can also be the case with live baits, for everything from panfish to catfish. In warm and stained waters, big live baits are going to create more disturbance underwater as they struggle against the hook. The bigger live baits often get eaten far more than smaller ones do in this situation. But smaller live baits often take lots of fish in clear waters, especially when it's cool.

Prevailing Winds

Not enough can be said about the wind. There are a great many benefits of wind, and some drawbacks as well. Most anglers hide from the wind, and this is almost always a colossal mistake. It's easy to understand why it happens. It's just simply a whole lot easier to run and hold a boat without being blasted by big wind and the waves that accompany it. But hiding from the wind means hiding from active fish, plain and simple.

I've seen it a thousand times. I head out on the water on a windy day, and the bulk of the anglers will have their boats in the quiet protected coves, bays, and little pockets out of the wind on one side of the lake. Most of the time, if you talk to any of those anglers at the marina or on the boat ramp lot at the end of the day, they'll say they didn't catch squat. We, on the other hand, likely had a great day of fishing. Why? Because we always fish structure and cover on the windblown side of the lake, no matter where we go. Windy days are almost always better days to fish than calm ones. But the majority of the benefits that an angler garners from fishing windy days is negated if they hide from the wind and fish calm areas.

Wind does a lot of things when it comes to fish and fishing. First, wind breaks up light rays entering the water. This allows predators like game fish to avoid being detected by their prey. Baitfish, crayfish, and other aquatic creatures that sport fish live on do not notice these hunters as well when the wind breaks up the surface, creating noise and scattering light rays down beneath. Wind also concentrates plankton, the microscopic organisms at the bottom of the food chain. Plankton float around in the water. After the wind has been blowing for a length of time, much of the plankton will get pushed up together into the wind-blown areas of the lake. Baitfish follow and eat the plankton. And we all know what eats the baitfish. The fish that we are after!

Game fish are predators that instinctively follow and use the wind to their advantage to hunt their unsuspecting prey. They take every advantage available to them, to hunt more effectively and to live in a more efficient manner. Living efficiently for sport fish means eating as much as possible while also expending the least amount of energy to do so. Like bodybuilders, predator fish will exert themselves for short periods and then swallow as much food down their throats as possible to get as big as they can be. These fish grow faster and grow to larger proportions. This means that they are bigger, healthier, and safer in their environment.

Not all winds are created equal. An old saying goes: Winds out of the east, fish bite the least, winds from the west, fishing is best. This time-tested piece of wisdom certainly rings true most of the time. Additionally, winds blowing from the south also tend to produce better fishing than those coming out of the north. But this is a lower effect than the east-west phenomenon. My favorite winds blow out of the west, but that is closely followed by winds from the south. North winds are next. North winds can produce excellent fishing at times, just not as consistently as those out of the west and south. I've seen however, subtle differences in this from time to time and place to place.

Lakes, ponds, and reservoirs tend to be more affected by wind direction than creeks and rivers. Oceans and seas are also far less affected by this than fresh waters seem to be. I've fished on waters out east that seem a little less impacted than fisheries are here in the central portion of the country. Regardless of where you are, you have to be able to deal with changes in wind direction. Most days, on most waters in Southern Illinois, the wind changing direction from west to east will dramatically slow, if not flat-out kill, the fishing. Changes from north to south are far less pronounced, but it will usually still alter fish behavior. This requires anglers to make adjustments to their game in order to get back on the fish.

Other weather conditions affect how much of an impact a change in wind direction can have. When the winds start blowing out of the east, some anglers on the water will become discouraged and head home. If they're not on the water yet, they might very well not go out at all. I've even had guided charter clients request to reschedule when the weather forecast shows winds out of the east for the day or days booked. I will admit that I never prefer to fish in east winds. And I have, when the schedule allows, switched trip dates for clients that want to change. However, as well everyone knows, you can't count on the weather forecast, especially a long-range forecast.

Meteorologists often get the wind directions and speeds pretty close within a few days. But a week or more out is a completely different thing altogether. I've had days that were supposed to be dominated by east winds that were out of the west or south instead, and the fishing has been incredible. Likewise, I've also seen plenty of days in which the winds were supposed to be optimal but the forecast changed several times before the trip, and the winds ended up blowing from the east. You can't count on the wind in the long-range forecast. Still, there are a number of factors that limit the ability of east winds to negatively impact a day of fishing. Likewise, there are a number of factors that combine with west and south winds that really crank the fish up and can lead to those trips that we dream of when we close our eyes at night. We'll get into those, as well as the rest of the conditions affecting fish behavior that anglers have to take into account when attempting to nail down effective patterns.

Cloud Cover

It's almost comical when someone who doesn't know much about fishing says to an angler, "Wow, I bet you wish you were fishing today, huh?" It's 80 degrees outside, bright, sunny, and clear as a bell, without a breath of wind. Uh, yeah, sure. Just like the fact that windy days almost always produce better fishing than calm ones, clouds are almost always preferable to sunshine. In fact, to a seasoned angler, the appearance of sunshine could almost be viewed as an appearance by the tax collector. Well maybe not that bad, but you get the idea.

Nasty-looking, dark, overcast days, when the sun never manages to shove its annoying head through, are quite often the days when you just cannot force yourself off the water. These can be the days that you fish until after dark, the days when your exhaustion isn't fully felt until well after you are at home and in bed, because of the adrenaline rush of excitement from all of the big fish catches that day. And when you finally do give in and call it quits, or rather some absurd aspect of life—you know like those ridiculous things called jobs—brings an end to your fishing trip for you, it's like one of those undisciplined kids being dragged from a toy store, kicking and screaming.

Yes, the days that nonanglers believe to be great for fishing seldom hold a candle to the ones they think we wouldn't be caught dead out in. I have to admit that I don't particularly care to fish in the rain, but how much I love those days when it looks like it should be raining. And the days when it appears as if a wicked storm is going to knock you for a loop any minute. And the days when it's so overcast that midday looks more like late evening. And the dark days that, well, you get the idea. I might have a bit of a problem. And it's been suggested that I have a love affair with the kind of weather that most folks run away from. The only exception to this rule would be in winter or very early spring, when waters are cold. Sunny conditions will warm waters faster. But fishing is still usually best when clouds move in, following days of sunshine that warm the shallows.

While cloud cover is icing on the cake of a day with good winds out of the south, the north, or especially the west, clouds can be a great equalizer as well. Clouds, and especially significant darker cloud cover, can save a poor day of fishing. I've had days in which there wasn't a breath of wind, or we had winds out of the east following a cold front when the bite was all but nonexistent, that were turned completely around with the arrival of cloud cover. I've seen clouds turn a water that seemed like the dead sea into a sportfishing paradise. From day to day, I'm always watching the cloud cover predicted in the weather forecast. Even on those rare days when I might not have otherwise fished, I'll often drop whatever I'm doing and get out there if significant cloud cover is expected.

Fronts and Storms

Storms are great for fishing too. When a storm front is approaching, the fish can sense it and feel it, as the winds pick up, the skies darken, and the barometric pressure changes. Fish will often go on a feeding binge that increases more and more as the storm front gets closer to the water. They often feed with reckless abandon, trying to fill them- selves to capacity before the storm front rolls through. And Southern Illinois anglers can capitalize on this feeding binge. Weather is usually warmer prior to the arrival of a front. And it often cools down, some- times considerably, after its passage through the area. This is referred to as the dreaded cold front. Most anglers with some knowledge about the sport know that the cold front usually means poor fishing. But it still amazes me how many people get the timing wrong. They often end up avoiding the water on the day that the storm front arrives, or even the windy day or days leading up to it, which are the best days to be on the water fishing.

When anglers refer to cold fronts in the context of slow fishing action, we are referring to the passage of a front, when a storm has already rolled through the area we're fishing, and left cool or cold air tem- peratures behind. This is generally the time to be avoided, if you can pick and choose your weekly fishing days that is. The postfrontal time frame is what we call the dreaded cold front in fishing. And this almost always produces poor conditions and the slowest fishing action. Baitfish schools often scatter and change locations, shallow fish that were active and aggressively feeding often slide down into deeper waters and get a case a lockjaw. Depending on the severity of the front, this period of low activity usually lasts for about two to four days.

This is nature that we are dealing with. So, there are no hard and fast rules that always ring true. But cold fronts will usually negatively impact fish and fishing more in winter and spring than in summer and fall. The impact is worse, or at least seems worse, in winter because the fishing action is already slow. This is a result of cold-water temperatures and the sluggishness of fish in frigid conditions, far outside their optimal temperature range, and with a lowered metabolism. The impact is also worse in spring, because this is a time when fish expect air and water temperatures to be warming, not going the other direction. They actively seek out warmer waters and get more aggressive as the waters warm. So significant cold fronts can reverse that process temporarily.

Summertime cold fronts are still not great. But the impact is far less than in winter and spring for a number of reasons, primarily because their metabolism is so high in the warm or even hot waters. Additionally, during periods of particularly hot and dry weather, a cold front can bring the water temperatures back down a little closer to the fish's ideal and preferred range. And the rain that comes with the front is beneficial, especially in those dog days.

Even though you never want to have to deal with a cold front at all, fall is typically the best time of year to have to do so. Fall fish are expecting the air and waters to cool down as the season progresses. So cold fronts don't slow the fishing down as much, or as long. Additionally, fall is a time when fish fatten up more than most other times of year. They feed heavily and often in an attempt to gain as much weight as possible before cold weather sets in, when they will feed less. This not only helps them reach maximum size, an advantage in health and safety, but also aids in egg development for the upcoming spring spawn. So, fish step up to the buffet and fill their plates as many times as they can.

While much of spring sees water temperatures at or close to the preferred optimal range of most game fish species, in spring fish are not singularly focused. Yes, they do feed a lot at times. But the spawn is the primary force that drives their behavior. This is so much so that both males and females drop significant weight through the rigors of procreation and can look skinny and beat-up for a while after the spawn has ended. In fall, though, water temperatures are at or close to optimal, and the main focus is feeding, feeding, and more feeding.

Stratification and Turnover

In summer, most waters in Southern Illinois experience stratification. Stratification can occur on rivers and streams at times, but it is far less pronounced than on lakes, reservoirs, and some large deep ponds and strip pits. Waters that remain more still will stratify easier than moving waters. But this situation doesn't affect river fish to a great extent. It does, however, considerably affect most fish in lakes. To put it simply, this is a process in which a body of water basically separates into three different sections based on water temperature. Cold water is denser, and warm water is lighter. As the high temperatures of summer considerably warm the shallows, cold water sinks, which occurs in areas of a lake with a depth of around ten to twelve feet or greater.

The warm upper layer of water is referred to as the epilimnion. The cold bottom layer of water is called the hypolimnion. And the middle layer of water, called the metalimnion, is where the water temperatures change quickly. The thermocline is located within the metalimnion. It's a thin band of water with the greatest change of temperature, which can be viewed with sonar equipment. One might simply assume that warm-water species of fish like bass, bluegills, and catfish would spend the summer living in the epilimnion, while cold-water species like muskies, trout, and walleyes would stay down in the hypolimnion where it remains cold, but it's not that simple. As the summer progresses, dissolved oxygen disappears from the hypolimnion. Most fish species cannot live in this zone, or at least cannot remain in the depths for long periods of time.

The wind keeps the warm epilimnion mixed well, and oxygen is usually high. The metalimnion, with its rapid temperature change, acts as a barrier to this mixing and prevents the hypolimnion from exchanging oxygen with the atmosphere. Additionally, aquatic vegetation does not grow in the dark depths. So, oxygen is not produced through photosynthesis either. As the season progresses, this can become a zone mostly devoid of life. While warm-water species have no trouble living and hunting in the warm or even hot waters of the epilimnion, some baitfish and some cold-water game fish species can become stressed during portions of middle and late summer. Many waters all over the world have experienced fish die-offs as a result of this. Sometimes it's significant enough to have major impacts on fisheries that are important for food and sport, seriously affecting local communities. While this is rare in Southern Illinois, it has happened here on a few waters.

Still, this doesn't usually occur except during extended periods of unusually hot and dry weather. Typically, it's been during the months of July and August, in the dog days. With what we've experienced over the last decade or so, I expect that these conditions may worsen and continue into September as well. September is often blisteringly hot now, and fish behavior is changing as a result, something that most serious anglers have already been noticing. Still, for right now anyway, most fish are able to regulate their body temperature while also getting enough oxygen to survive. However, during those particularly long and hot summers, some fish will venture into the cold hypolimnion to bring their body temperature down before coming back up into the warmer, highly oxygenated and prey-rich waters. It's a balancing act of sorts that places additional stress on these animals. It is for these reasons that most local muskie anglers do not fish for muskies in the heat of summer.

Some species of game fish are pursued by most anglers primarily for sport, while others are pursued primarily for food. Most bass anglers, for instance, fish primarily for the sport of it. Therefore, most carefully release all of the bass they catch. The majority of anglers who target crappie, on the other hand, do so primarily for food. They usually keep all the legal-sized crappie that they catch within the limit. Like bass, muskies are considered a top game fish. Nearly all anglers who pursue them do so exclusively for sport. Like bass anglers, muskie anglers also tend to be very conservation minded, so much so that they study the latest release methods to help ensure the survival of the fish that they catch and carefully return to the water.

Southern Illinois muskie anglers generally advise against targeting this species in the middle of summer. Luckily, hooking muskies in extreme heat is rare. But as previously mentioned, when this does happen, we try to land and release the fish as fast as possible. We also never remove it from the water. These same release techniques apply to walleyes and trout. These fish also become easily stressed in summer. Although most anglers targeting trout and walleyes keep the fish they catch. The striped bass is another big-game species like the muskie that becomes stressed in hot waters. Stripers prefer cooler temperatures. While it is uncommon to hook them in the heat of summer, the same release techniques used for muskies work on stripers. Generally, larger and older fish become stressed more easily. While this time frame might not cause too many problems for small muskies, stripers, trout, or walleyes, the bigger specimens can't handle the same difficulties.

The positive aspect of fishing when lakes are stratified, is that *deep* becomes a relative term. Active fish are more concentrated. The thermocline will move, and the metalimnion can become deeper or shallower from day to day and week to week. It can even change depth throughout the day. Some fish can still use deep water a bit. But so much of the depths are eliminated that it does make fish easier to find and target. By locating the depth of the thermocline, an angler can determine what depths are productive from the surface down to the level of the thermocline. Some fish will spend more time near the surface, while others might spend most of the time at the maximum depth containing adequate oxygen content. But the majority will move around in the constant search for food.

The depth of the thermocline depends upon many factors. In general, deep lakes have deeper thermoclines, and shallow lakes have shallower thermoclines. Additionally, water clarity plays a major role. Clear waters tend to have the deepest thermoclines, while heavily stained waters have shallower thermoclines. Personally, I've marked thermoclines as shallow as about 11 or 12 feet and as deep as 30 on Southern Illinois waters. Thermoclines tend to get a bit deeper as the summer progresses.

The best way to locate the depth of the thermocline is with a sonar unit. Sonar equipment has been commonplace on most powerboats and nearly all fishing vessels for decades. But in recent years, many ice anglers have used portable units for hard water fishing. This has led to these portable units being employed on small paddling craft, as well as those powered exclusively with electric trolling motors. Anglers now use portable units when fishing from canoes, rafts, belly boats, and other small craft. Since the sport of kayak fishing has been growing so fast in popularity, all kinds of gear have been added to them. Anglers use portable sonar equipment on many styles of kayaks. But some anglers permanently mount sonar units on some of the fishing design yaks, along with trolling motors, rod holders, miniature live wells, and other accessories. The small craft have become less expensive options for anglers than larger gasoline-engine-powered boats to target deeper waters further offshore.

Whether fishing from a small kayak or a big bass boat, sonar shows us where to fish. The thermocline will appear as a thin unending line of clutter on the sonar screen. If an angler can't find the thermocline with a unit on factory settings, the sensitivity can be increased to pick up the temperature change that is so important to fish. Just refer to the owner's manual of the particular unit. Scuba divers and free divers can also feel the thermocline easily. But for anglers without access to any kind of boat and sonar unit, or a friend who dives, a call to a local IDNR fisheries biologist might provide the information one needs to more effectively target fish from bluffs, bridges, and deep, steep sloping shoreline banks.

Turnover occurs in fall, after shallow waters cool considerably and begin to sink. In this instance, usually with some significant wind, the waters will begin to mix from top to bottom. The thermocline disappears and cannot be marked by sonar, and then the water will usually become more stained, almost dark for a period of time. It's common to see all kinds of gunk (highly technical term, I know) floating on the surface and suspended in the water column. This is from decayed leaves, weeds, and other plant material coming up off the bottom during the turnover.

Many people refer to turnover like the dreaded cold front. Turnover does change things in the fish's environment. And fish don't care for change all that much. This phenomenon does affect fish behavior in a negative way, and it is common for fishing action to drop off for a period of time as a result. But it's not the boogeyman that many anglers make it out to be, and there are some things we can do to offset this and stay on biting fish. By far the biggest factors I've found for catching fish during and shortly after turnover are going super shallow and moving away from expanses of deeper water. These aren't actually huge moves anyway, since anglers should already be focused on fishing shallow structures in confined areas. But fishing extremely shallow waters, and just as importantly, fishing confined locations that are far away from water that will be turning over, becomes paramount.

Get far away from the main lake basin, and also far away from deeper sections of creek arms and bays. The idea is to find very shallow areas, way up in the back portions of long creek arms and big coves, where you have shallow water extending a good distance out from the super shallow water. When you locate an area like this, check the creek channel. If the creek arm has a well-defined channel that's over about ten feet, then move away from that channel as well. Ideally, you'll fish areas with shallower creek channels. But if not, the best arms are larger ones where the creek channel is off on one side of the arm rather than running down the middle. Then, by moving across the large creek arm, anglers are able to distance themselves even more from any water that might be turning over.

In some Southern Illinois waters, coves can be better options than creeks if a lake has creeks that are too deep. Coves don't usually have quite as large of an expanse of shallow water. But without a significant creek coming in the back, they won't have a defined channel that might place deep water in the middle of an otherwise shallow-water fall fishing paradise. Using a lake map with depth contours is the quickest way to search out prime locations. Once in the area, go through and find the deepest spots of the cove or the channel of a creek. Follow them into the shallows, playing close attention to the sonar screen to verify the depth. It can take a bit of time to search out and learn how to fish new areas like these. But it's something that will pay off big time when turnover comes to town.

Fishing Pressure

I've advised my clients the same thing for two decades now. If they can possibly swing it, it is far better to book guided charter trips during the week instead of on the weekends, when most people are off work. It is best to go fishing whenever you can go. But anglers fishing during the week possess a huge advantage over those fishing weekends. Weather conditions do affect fish behavior more than things like boat traffic and fishing pressure. But man-created influences do play a major role. Saturday and Sunday are usually about equally poor days to fish from a pressure standpoint. Saturdays usually see a few more tournaments and a bit more boat traffic than Sundays do. But Sunday is the second day that the water and the fish have been beaten up on, so they're raw by this time.

With the exception of holidays, Mondays are usually pretty good, but not optimal. Monday mornings can sometimes be a little tough after a busy weekend. The fish might still be in the process of settling down, and the water might still be in the process of clearing back up. Pressure starts picking up again on Friday, especially in the second half of the day. In order of importance, and excluding holidays, the best days to fish are usually Wednesday, Tuesday, Thursday, Monday, Friday, Sunday, and Saturday, in that order.

When fish are constantly being fished, like any creature, they get wise to it. And they learn to avoid capture. Just the mere fact that a dozen other anglers may have come down the point that you're now fishing later in the day can lead to no results for you. Of course, anglers taking fish home for the grill or the frying pan means fewer fish that you could potentially catch on your favorite structure. But even with catch and release, provided it's a good and healthy release that is, the fish might still move off to another location. Even if they remain where they are, they'll be unlikely to bite again for some time after being caught. All of this reduces the number of catchable fish remaining and creates tougher angling.

Even when SI fish don't bite at all, they're still less likely to take your offering after they've had a half dozen others bounced in front of their faces before you got to them. All of this, with boats passing over, noise from outboards and trolling motors, people talking and making other sounds that carry into the water, lures splashing down on the cast, and so on. Fishing pressure can really change things. Focusing the majority of one's fishing time during the week makes a big difference in success. But if we're stuck fishing mostly weekends, what do we do? Sometimes a switch from artificial lures to live baits can put more fish in the boat, especially when large numbers of fish are concentrated and water coverage is not an issue.

Provided the water is clear enough, downsizing to smaller lures can sometimes trip the trigger of fish that have shunned standard-sized offerings all day long. Switching to quieter and more subtle lures that displace less water and create reduced vibration can sometimes be a hot a ticket to success. Slowing down the retrieve speed and teasing the fish into biting can sometimes work. On the other end of the spectrum, speeding up considerably and giving the fish a short look at a lure can sometimes cause their predatory instincts to take over and get them to attack. If the weather conditions are good, sometimes fishing pressure has little impact. But when it does, be ready and willing to pull everything out of your bag of tricks.

Boat Traffic

Everyone talks about fishing pressure. There is no doubt that periods of high fishing pressure do impact fish in a negative way, and much more so than boat traffic. But all use of the water by humans can and will affect fish behavior. Boats are noisy things that make a lot of racket and churn up the water. There is no question about that. High recreational boat traffic can absolutely change fish behavior and slow the fishing action down all by itself. Summer is a prime example. Not only do larger lakes like Crab Orchard, Kinkaid, and Rend have anglers using them, but they see additional traffic from houseboats, cruisers, pontoons, sailboats, ski boats, jet skis and other craft. These popular lakes even have party coves on them, and most smaller waters can see less boat traffic per acre on summer weekends. Fishing is practiced much less in summer than in spring. Most anglers consider spring to be the focal point of the fishing year. There is no question that more fish and bigger fish are caught during that time than any other, at least in our neck of the woods anyway. But there are also a lot more anglers on the water.

Besides weather that may be a little too hot, most knowledgeable and experienced anglers know that spring fishing is far better than it is during most of summer. They focus so much time, vacation days, and money on spring fishing trips that by the time the heat of summer sets in, most don't fish nearly as often. While fishing by novices increases, with families vacationing with the kids, the serious anglers have already beaten the water to a froth for months. Serious anglers are more likely in the summer to only make the occasional day out to the lake or river in the heat, here and there. Fishing pressure definitely drops, yet fish still behave the same way on summer weekends as they did in spring when there was a lot more fishing pressure but much less boat traffic and overall water use. This is proof that boat traffic alone can have a significant negative impact on fish and fishing action.

I can't count how many times the fish were on fire all week long but slowed considerably by Friday evening and becoming nonexistent by midday on Saturday. The shallows often get muddied up, and I've seen that throw the fish off. Sure, sometimes you can switch to bigger presentations or go to dark or bright colors that stand out more. You can also switch to rattling lures with more vibration or put out huge live baits. But that doesn't always work. Changing locations is next. You try to find fish in other areas that aren't quite as churned up. But whether you can find enough fish that are also active becomes another story. Sometimes adjustments work, but other times the fish get lockjaw for the rest of the weekend.

One of my pet peeves is when people drive through shallow weed beds, especially when I'm on a hot weed bite. I'm always amazed at just how many people who don't know a lake well enough will tear through shallow water, just hoping not to hit anything. I've certainly seen it happen. I've witnessed boaters destroy lower units, tear out-boards right off the backs of boats, and rip holes through the bottom of aluminum and fiberglass hulls. These are the kinds of boaters who take a red-hot weed bed and ruin it for days by overly chopping the salad. Fish scatter or just stop biting. But regardless, it's time to start working on a new pattern again. Yes, boat traffic can be a major ob-stacle for the angler to climb.

Solunar Peaks

The sun and the moon not only affect the planet and all of its cycles, but they also affect all living creatures on Earth. Humans are of course included in this. But we just don't feel or understand the effects on us in the way that our ancestors did. Hundreds of years of disconnection to nature, through civilization, has dumbed us down and dulled our senses. And ridiculously, it's even caused some of us to see ourselves not as a part of nature but as separate from it. Luckily, as mentioned previously, with time and effort we can place some distance between us and society and develop our senses as we build a stronger bond with the natural world.

Fish and game are absolutely affected by everything in their world, including light and gravitational pull. Anglers and hunters can access solunar charts. Based on the cycles of the sun and moon, and on your location (latitude and longitude), these determine days of every month, as well as A.M. and P.M. times of each day, in which the sun and moon can cause increases in fish and game activity. These solunar peak days and peak times offer a better chance to catch fish or see animals. Again, these do not affect fish behavior and angler catch rates as much as the weather does. Weather is always going to be the biggest factor in fish activity and angler success from a standpoint of conditions. But solunar peaks do play a role.

The full moon and new moon phases create the best opportunities each month. Typically, the day of the new moon and the day of the full moon are the two best days to fish each month, from the standpoint of the solunar cycles. The three days prior to each of these peak days, as well as the three days following each peak day, are also considered optimal days of the month for fishing. Generally, fish activity is more likely the closer each day is to the new or full moon. But the days prior to each phase are often considered a bit stronger than the days following each phase. Additionally, while of less importance, the two quarter-moon days of each month also offer a minor peak as well. Based on the charts, there are also two peak periods of time each day, one in the first half of the day, and one in the second half. All of these peaks often produce short but intense feeding windows that Southern Illinois anglers can capitalize on as fish activity increases.

Other times to consider each day are sunrise and sunset, and moonrise and moonset, which can and often do increase fish activity and feeding. And when moonrise or moonset happen to fall close to either sunrise or sunset, that feeding window can be even more intense. I've noticed all of these cycle peaks to ring true. And I cannot count the number of times that a slow day of fishing turned red hot during one of these daily peak periods. However, it baffles me just how many anglers fail to factor these peaks into their fishing game and miss out on opportunities. Factoring these in each day means trying to make sure to get out on the water in time to hit one or more of these peaks. It also means trying to schedule fishing days on or close to the peak period days of the month. Little things like this can add up over time, in both numbers and size of fish caught.

Prime Times

Conditions for fishing are different from time period to time period throughout the day. All fish and game behave differently at different times. When it comes to freshwater fish in general and all of the game fish species found in Southern Illinois, increases in activity happen most frequently during the low light periods throughout most of the year. And taking this into consideration can lead to greater success on the water. The conditions of early morning and late evening lend well to catching fish, and especially to catching big fish.

While I've certainly had great fishing in the bottom of the Prairie State through the heat of the day, it's the exception rather than the rule. Except for the cold-water period, the best fishing all year long revolves around morning and evening. Sure, we do catch fish during the middle of the day and in early afternoon throughout spring and fall. But we catch far more fish, and much bigger fish, early and late in the day during these seasons. This is even more the case in summer. High heat and bright sunlight usually don't do the angler any favors. It slows down fish activity and significantly reduces catch rates. The only exception being on those dark and cloudy or very windy days, during or prior to storm fronts.

In summer, it's best to be on a top spot before light and then fish until late morning. The bite usually slows considerably through midday and all of the afternoon. It usually doesn't pick back up until at least the middle of the evening. Fishing the shady spots as long as possible in the morning, and again as they become present in evening, is a trick that puts a lot more fish and bigger fish in the boat for at least half the year. But it's one that most anglers fail to pay attention to. Winter, and the very early portion of spring, are the only times of year when I do like to fish through midday and the early afternoon. Because the waters are so cold, the sunshine at this time of day warms them up and often produces the only really hot bite I'll get.

Night Moves

Night fishing is an option that few Southern Illinois anglers take advantage of. And the radically different conditions can offer advantages at times. As the years have passed, I have done more and more night fishing, and it's paid huge dividends. While I could count the number of times on my hands that I've night fished in winter, it's likely a good bet. Especially early in the night, on one of those sunny days that warmed the waters a bit. Like most, I'll admit that the cold primarily keeps me from it. But it's a good idea for anyone that can stick it out. Water clarity is typically the best from late summer through the middle of winter. Clear water is important for consistent night fishing success for most species, and especially with the use of artificial lures. So, night fishing from August through January would usually offer the best conditions. Still, clarity plays such an important role for most nighttime anglers that lake choice can be critical. Fishing the clearer waters of Southern Illinois is your best bet for hot nighttime fishing action with most species. The only major species exception here would be catfish.

Catfish utilize their sense of smell much more than bass, crappie, stripers, carp, and most species do when hunting. And while cats can be caught on artificial lures at times, most primarily take live and dead natural baits. When targeting the various catfishes exclusively, both clear and stained waters produce well at night. And monsters have been taken in the dark, even on the muddiest lakes and rivers. For just about every other predator however, and certainly most of the game fish species other than catfish, clearer water is the magic ticket to success long after the sun has fallen below the tree line. While the typical heavier rainfalls from late winter to early summer keep most waters muddier, a hard rain can temporarily cast a heavy stain to any potential night fishing lake at any time of the year. So, it's best to stay on top of the weather and water conditions before heading out after dark.

Night fishing can be great in the summer here, when it's too hot to be out for much of the day. Besides the more pleasant temperatures to be fishing in, fish activity is often higher at night than it is during the heat of the day. As discussed previously, early morning and very late evening are the best time periods during the day. But often, the evening bite can be very short. Good night fishing can sometimes be had all summer long, and this can be a great option. Fall is another excellent time to night fish. While the wet weather of spring often muddies the waters up too much for successful night fishing for any species except catfish, fall usually offers great night fishing for all species. The clear waters and cooling temperatures that are common in fall mix up a great recipe for awesome night fishing. And with increased fish activity, they will often feed all night long. Another huge benefit to fishing at night is the reduction in pressure. Anglers and pleasure boaters largely abandon the waters at night, fish settle down, and you're likely to have the top spots all to yourself.

Perfection

Sometimes the sun, the moon, the stars, and everything else line up just right. And it's those days that no amount of money could ever buy. Those days are immeasurable, the kind of experiences that produce invaluable memories. Dad and I were fishing together on Kinkaid when we experienced the fastest muskie action that either one of us has ever had. The fishing was a little bit slow earlier that day. But the wind was blowing, and conditions were changing. While working across a large flat that had been loaded with shad for weeks, Dad hooked up and landed a quality muskie. Immediately after we released Dad's fish, I hooked up on another similar-sized muskie.

Dad scooped her up in the net. And as we were unhooking and releasing the fish, we noticed a number of blow-ups on the surface, game fish attacking prey on top nearby. We dropped the Power Pole shallow-water anchor in order to hold the boat perfectly still in the middle of the flat, and we never moved. In less than an hour of time, we hooked, fought, landed, and released an almost unbelievable seven muskies! Seven muskies boated in an hour, in addition to other follows and strikes. It was nothing short of amazing. The season of the year had the fish bunched up. And with concentrations of baitfish, the wind, and the approaching front, they just went nuts. Once fish number two was hooked, and with other blow-ups nearby, we decided to land and release each fish as fast as possible. We wanted to take full advantage of the feeding window we were experiencing, and that we did. It's unlikely something like this will ever happen for us again, but what a rush!

7

Prey and Hunting Methods

There's an old term called *match the hatch*. Fly anglers came up with this. It basically refers to insect hatches, and how fly fishers must often be able to pull a fly out of their box that very closely resembles the particular insect that happens to hatch at a given time. Stream trout are particularly known for being picky with what they eat. And once they begin feeding on a specific type of bug, they may shun everything else for a time. Matching the hatch does cross over to most other species of game fish here in Southern Illinois, and even some rough fish too. Everything from bluegills to bass and crappies to carp has a favorite prey item, but they also get used to feeding on a particular type of prey that becomes more readily available at certain times.

Match That Hatch

This patterning trick is not just related to insects either. Matching the hatch could mean using a crawfish imitating crankbait for smallmouth bass in a creek, or a topwater frog for farm pond largemouth. It could mean using a multiblade spinnerbait that mimics schooling shad for muskies or stripers on a lake where they most commonly prey upon gizzard shad. Or it could be using live green sunfish for flatheads and channel catfish in a river with a high population of greenies. Matching the hatch means getting close with various factors such as size, shape, and color of prey items. Sometimes, various conditions require the opposite approach. But it's usually best to begin by trying to match the hatch first, and then make adjustments as needed.

Available prey is common knowledge on some waters in our region here. But on those where it's not, this information can be obtained by talking with other anglers you know who fish the water regularly, by chartering a trip with a fishing guide, and by talking with a local, state, or federal fisheries biologist for that area. Before we get into the specific patterns themselves, it's important to note the various prey species most commonly utilized by Southern Illinois's aquatic predators, as well as where the various predators prefer to hunt this prey. This gives us a better starting point when selecting everything, from lures to lines to rods, to put together the most effective patterns for our primary target species.

Common Prey Species

Various species of shad, shiners, herring, suckers, chubs, minnows, and other similar baitfish often make up the majority, or at least a high volume, of the diet of most game fish on most of the larger lakes and reservoirs, and also on the big rivers and larger feeder streams of Southern Illinois. Where available and abundant, these kinds of baitfish are often the preferred food source of adult SI predators, for several reasons. Many of these species are pelagic, schooling together in numbers and roaming around a water in search of food. As a result, predator species can follow these schools around, making multiple predations on the unlucky individuals over a length of time. It's similar to the buffet style of restaurant for humans.

Additionally, these baitfish have fatty and oily flesh. In much the same way that bodybuilders and powerlifters turn to higher calorie foods and weight gainer supplements during portions of their training seasons, many game fish will actively seek out these oily baitfish species where available. They'll prey on them more than other items because the calorie content of these fish is much higher than others. They get more bang for the buck, so to speak, maximizing size and health as they ingest more calories with less effort expended.

Bluegills, pumpkinseeds, warmouth, and other various sunfish species often make up the majority of the diet of game fish in smaller waters. Lakes without pelagic schooling baitfish like shad, herring, and shiners will often produce far more fish on lures that imitate bluegills rather than those which mimic threadfin shad. Predators are, for the most part, opportunistic. They will often readily attack whatever is careless enough to get within range of an easy strike. So, bass on large lakes do eat both shad and sunfish. It's just that where available, most predators will choose shad over sunfish. This isn't always the case though.

Prey Preference

Take for instance the difference between two top predators, the muskie and the flathead catfish. These are both large and powerful animals, apex predators at the top of the food chain, like lions or orcas. Of the various Southern Illinois waters that harbor muskies or flatheads, the best place to catch muskies would be Kinkaid Lake, while the top place for flatheads would be the Mississippi River, but both waters contain the favored prey species of both super predators. Both of these game fish feed on shad and sunfish in all places where both prey species are found in abundance. But muskies will usually feed more on shad, while flatheads will usually feed more on sunfish. And this has to do with where the fish live and how they hunt.

Muskies are a species that tend to feed up. They spend a lot of their hunting time closer to the surface than flathead catfish. Flathead catfish spend most of their lives on or close to the bottom, where muskies don't feed as much. Schooling baitfish like shad are always on the move. They don't take up residence in specific places and don't typically hold in cover. Shad tend to suspend in the water column, whereas sunfishes spend a lot of time holding along the bottom. Bluegills and other sunfish species also use cover where available. Because of these major differences in where these two predators live and how they hunt, it becomes easy to see why muskies eat more shad where available, while flatheads eat more sunfish instead.

Another prime example of this would be the striped bass and the common carp. Common carp are pretty much exclusively bottom feeders, even more so than flatheads. Striped bass, on the other hand, feed up like muskies, or even a little more so. As a result, prey species such as earthworms and crayfish are popular in the diet of carp, while shiners and herring are top options for stripers. Since we touched on carp, we'll mention a few other rough fish quickly. Drum, bowfin, and the various species of gar can be caught on artificial lures, even more so than catfish. But they all take various live and dead baitfish and panfish too.

They also eat crayfish, worms and aquatic insects. Chubs and suckers are a sort of in-between kind of baitfish. These species do school up in groups and move around a lot, but they tend to spend more time on and near bottom. These are perfect prey for flatheads but are not nearly as common as sunfishes in most waters. Some anglers target these, like they would larger rough fish. And these fish can be caught on worms, insects and small crayfish. We've even landed a few on small minnows fished for crappie, and they make awesome flathead baits.

Small Southern Illinois game fish species such as sunfish, crappie, and bullhead catfish are utilized regularly by larger predators. But all of these predators are cannibalistic too and will readily attack their own kind. Young of the year game fish, especially the smaller species we've covered, as well as middle-sized species, like all of the various black basses, largemouth, smallmouth, and spotted bass; the smaller temperate bass species like white bass, yellow bass, and the hybrids; plus walleyes, saugers, rainbow trout, and channel catfish are regularly preyed upon as fingerlings and yearlings. Larger species like pure stripers, muskies, and flathead and blue catfish do get eaten when young, but in fewer numbers because of their larger sizes. All of these species are more likely to be preyed upon in waters lacking oily schooling baitfish though. All of the larger lakes and reservoirs like Cedar, Crab, Egypt, Kinkaid, Newton, and Rend harbor at least one type of schooling baitfish species like shad, as do most medium-sized lakes like East Fork, Forbes, Grassy, and Kitchen. All of the major rivers and the larger creeks that feed into them contain multiple types of these oily pelagic prey, including various species of shad, shiners, herring, and other minnows. It is the smaller lakes, creeks, and ponds that are far less likely to hold baitfish. In these waters, bluegills and other sunfish species are top prey items, and all young game fish are more susceptible to predation.

Still, with larger game fish, much more is possible. A 3-pound trophy-sized crappie, or even a carp that's double that size, doesn't look very big in the jaws of a muskie or flathead tipping the scales at 25 or 30 pounds. And these fish grow larger still. What is too big? An 8-pound fish, or maybe 10 or 12? It's hard to say. I can't count how many times over the years that I've caught fish of many different species, on lures and baits that were as big as they were or even larger! I landed an average-sized Kinkaid crappie with its mouth pinned open by the hook of the muskie spinnerbait it attacked, and I caught a Mississippi sauger on a minnow bait that was longer than the sauger. The gaping maw of a largemouth bass is absolutely enormous for its length and weight. Largemouths are capable of waging an effective assault on truly huge prey.

Smallmouth bass of course have much smaller mouths for their body size, as the name suggests. But the aggressive nature of the smallmouth never ceases to amaze. We've landed bronzebacks of all sizes on lures that appear much too large for them to swallow. Yet they savagely maul them anyway. Many of the biggest smallmouth bass that Dad and Amber and I have caught in the northern United States and Canada have come on large lures intended for northern pike, muskies, or lake trout. I'm talking artificial baits in the 6-to-10-inch range, weighing anywhere from 1 to 5 ounces! Of course, that scenario is less common in Illinois. But we have taken many smallmouth bass here on oversized lures. The same is true with walleyes and sometimes even channel catfish.

Many folks remember the video floating around the internet years ago of the big flathead catfish that was swimming around the surface of a lake. It was stuck, and unable to close its jaws or swim below the surface, because it attempted to eat a kickball! Yes, for those who didn't see it, the catfish apparently attacked and attempted to swallow a full-size inflated red kickball and required the help of boaters passing by. Dad landed a super-fat 48-inch beast of a late-fall SI muskie on one of our Hatchet Shad spinnerbaits that regurgitated a squirrel in net! The Hatchet Shad is a pretty big bait. This particular model, one of 3 sizes we designed, stretches the tape to 7.5 inches long and weighs in at 1-1/2 ounces. With 3 blades, a bulky skirt, and a swimbait tail, it's a fairly large lure. This gluttonous muskie inhaled the big lure after having eaten what appeared to be shad that it coughed up, in addition to the large adult squirrel.

While small game fish like bluegills, pumpkinseeds, and redear sunfish usually take only small lures and baits that will fit into their tiny mouths, we occasionally catch them on bigger lures. Often, though, they're hooked on the outside of their mouths. They try, but their jaws just won't accommodate what their instincts trigger them to attempt to eat. Green sunfish, warmouths, and rock bass on the other hand are on the opposite end of the sunfish spectrum. These species actually have much larger mouths for their body size. And they readily use them to their advantage to tackle good-sized prey.

The mouth of a green sunfish, a warmouth, or a rock bass more closely resembles that of a largemouth bass than a bluegill or redear. In fact, in body shape, and with their oversized jaws, they're built almost exactly like most ocean grouper species. They're very aggressive and attack huge prey for their size, just like grouper will. About 100 miles offshore, in international waters near the Texas-Mexico line, I once landed a massive grouper on a *Field & Stream Hook Shots* show with Joe Cermele. The fish engulfed the biggest bait we had on board that day. But its belly was already so full that it was swollen, and you could feel hard objects through the skin, big freshly chomped prey. The bigger-mouthed sunnies here in Southern Illinois behave the same way that bigger-mouthed groupers do down south. Nature is truly amazing! But I've said it time and again that if green sunfish, rock bass, or warmouth grew to the size of muskies or stripers, or even largemouth bass, few anglers would target any other species. It would just be too much fun chasing these brutes. Actually, on light tackle, they are a heck of a lot of fun to catch.

Almost all fish feed on other fish. With the exception of filter feeders like paddlefish, all of the game fish species and most of the rough fish species will feed on pretty much all of the baitfish species when they are available. Most Southern Illinois game fish and many rough fish will also feed on sunfishes, as well as other baby and juvenile game fish and rough fish species, just with various preferences like we've discussed. However, some fish prefer to eat various fish alive while others prefer fish that are dead. Most game fish, all of the basses, black and temperate, walleyes, saugers and hybrids, trout, crappies, bluegills and other small sunfishes, and muskies all prefer live baitfish. That's not to say that these fish cannot be caught on a dead baitfish, because some occasionally are. But it is a rare occurrence.

Fish like walleyes, saugers, saugeyes, rainbow trout, bluegills, pumpkinseeds, and redear sunfish often take artificial lures and natural baits equally well. Most of the time, however, it is more efficient and more effective to fish for muskies, largemouth bass, smallmouths and spots, white and black crappies, green sunfish, rock bass, warmouths, and white and yellow bass with artificial lures rather than natural baits. With most species though, the key to using live baits is to use lively baits. You want good healthy live baits that move a lot on the hook. The ones that get nervous when a predator comes close and tries to evade it, triggering the predator fish's attack response, like a cat on a mouse.

Of course, when talking about prey, fish aren't the only things on the menu. Crayfish (freshwater crustaceans), also called crawfish, crawdads, or mudbugs, resemble tiny lobsters and are an abundant prey source in Southern Illinois and most areas throughout North America. Most game fish and rough fish prey heavily on crayfish where they are abundant. Frogs, and to a lesser extent toads, are eaten heavily by all the black basses, but especially largemouth bass. Bucketmouths just love frogs! Muskies also eat frogs frequently, and the occasional frog ends up down the throat of a flathead or channel catfish.

Rats, mice, muskrats, chipmunks, mink, and, as we've found, squirrels are on the menu for a variety of fish. When it comes to larger mammals, baby raccoons, opossums, otters, beavers, and similar furry critters could all end up as lunch. Muskies and largemouth bass would be the most likely culprits, but big gar, channel or flathead catfish, or even a striped bass could take a stab. Snakes are preyed upon as well, especially by largemouth bass and muskies, as are salamanders. Birds make for another possible meal for largemouth bass, muskies, stripers, catfish or other large predators that will feed on the surface and might get lucky.

I had a big SI channel cat cough up a bird one time that appeared to be a killdeer. All kinds of birds have been plucked from the limbs of trees or the water's edge. And some diving water birds get taken subsurface too. In spring, when ducks and geese have large numbers of young, top predators will sometimes pick off some of those at the edges of the group, especially the stragglers that don't remain close to the adults. Panfish and trout eat a lot of aquatic insects, but all of the black basses, white and yellow bass, walleyes, saugers and hybrids, and channel and bullhead catfish eat both aquatic and terrestrial insects as well.

Catfish are quite different than a lot of other game fish as a group, and they are also vastly different from species to species. All catfish species will eat dead prey. Flatheads definitely prefer live baits, very fresh, lively and active baits in fact. And from my experience, the flathead cats here in Southern Illinois seem to favor live baits maybe even a little more than those in other areas, especially on the large lakes and rivers. During most of the year, flatheads will not only shun dead baits but will even turn down an easy meal from a baitfish that isn't highly active. The biggest exception that exists is during winter and early spring, when catfish just start getting active again after the long cold period, with reduced feeding. At this time their stomachs have actually shrunk a bit, and they're not as willing to tackle the big lively prey they do during most of the year. Small live baits can work. But smaller-sized whole dead baitfish, or cut sections of baitfish, commonly just referred to as cut bait, are often preferred as the flathead's stomach adjusts to larger meals and more regular feeding again. The second time this can happen is for a short period of time following the spawn, when flatheads feed little or not at all, and again they often take a bit of time to adjust to big regular meals.

For the majority of the year however, the livelier the fish used for bait is, the more and bigger flatheads an angler is likely to catch. Flatheads can sometimes be taken on artificial lures too. While they usually prefer fish, they will also take other natural baits. Crayfish can work well, especially lively crayfish. But the majority of the catch will usually be smaller-sized flatheads. Young flatheads often prefer the mini lobsters when they are juveniles. But once they reach adulthood, they will almost always choose a fish. Night crawlers, red worms, grubs, turtles, and similar live natural items do produce. But most of the year it's primarily going to be small flats at the end of the line. Adult flatheads are apex predators, and live fish is their version of our filet mignon.

Blue catfish are at the opposite end of the spectrum. SI blue cats can sometimes be taken on live baits, but it's rare. Whole dead baitfish work well, but cut bait is the real ticket for catching adult blues. Cut bluegill or other sunfish will take blue cats, but oily baitfish such as gizzard or threadfin shad, skipjack herring, mooneye, golden shiners, chubs, and suckers produce the best cut bait for the blue catfish, since this is what they primarily feed on most of the year. I've never personally caught a blue cat on an artificial lure. And I don't believe I've ever spoken to someone who has. I've read about the incidental catch of a blue cat on a crankbait or a jig and plastic, probably doused in scent and fished extremely slowly in the water column, but blues would be the last catfish you'd ever hope to catch on anything artificial. They do eat crayfish, worms, and similar live and dead natural baits, but like with flatheads, it's mostly smaller fish.

Channel catfish are the most versatile in their feeding habits; that is to say, they are less picky than either flatheads or blue cats. Channels are also the catfish species that is most commonly caught on artificial offerings, and quite a few are taken each year with lures here. In fact, I've caught more channels on artificial lures in Southern Illinois than anywhere else. Crab, Grassy, Kinkaid, and Rend Lakes, and the Big Muddy, Mississippi, and Ohio Rivers cough up a lot of channels on lures. Channel catfish readily take any and all live and dead natural baits. They eat a lot of baitfish and panfish, both alive and dead, and take cut baits well too. They eat live and dead crustaceans, bugs, frogs and toads, all kinds of worms, aquatic insects, birds, you name it. They will eat from dead carcasses lying on the bottom, such as turtles or mammals. Whatever an SI channel cat can catch, it will eat. But their preference does change a bit with age.

Smaller channels mostly seem to prefer dead offerings, while larger channel cats definitely prefer live creatures. You can catch good numbers of these fish, of all sizes, on just about any natural bait. Cut bait or dead items often produce the fastest action. But the majority of the big channel catfish we've landed over the years have come on fresh lively and highly active medium-sized baitfish or panfish. Understanding the available prey items, which items are used by the various sport-fish species, and the interactions between these species (the predator-prey relationship) greatly helps an angler to effectively pattern the fish.

A peaceful morning for anglers in the swampy river lowlands of Southern Illinois. *CSO Team / Colby Simms Outdoors LLC*

The awesome power of a trophy-caliber Southern Illinois muskie is on full display during release. *CSO Team / Colby Simms Outdoors LLC*

Internationally renowned fishing pro Ray Simms with a monster SI bass on a CST Flash N Spin. *CSO Team / Colby Simms Outdoors LLC*

Big white and black crappie are available year-round in the Lower Land of Lincoln. *CSO Team / Colby Simms Outdoors LLC*

CSO Team's Shay Simms goes toe to toe in an epic battle with a trophy-class Southern Illinois catfish. *CSO Team / Colby Simms Outdoors LLC*

IDNR Fisheries biologist Shawn Hirst displays two largemouth bass during a lake sampling. *CSO Team / Colby Simms Outdoors LLC*

Author Colby Simms displays the razor-sharp teeth of this Southern Illinois muskie. *CSO Team / Colby Simms Outdoors LLC*

BWA TV host Mark Davis lands a big walleye on a shad pattern swimbait. *Big Water Adventures / DL Ventures LLC*

Huge rough fish like this Southern Illinois buffalo landed by the author offer tremendous sport. *CSO Team / Colby Simms Outdoors LLC*

Big lures like this Monster School N Shad spinnerbait catch trophy fish of many species. *CSO Team / Colby Simms Outdoors LLC*

Famous media personality and guide Ray Simms with a massive SI muskie on a CST Hatchet Shad. *CSO Team / Colby Simms Outdoors LLC*

The scenery of Southern Illinois's Ozark Mountain waters is unmatched and cherished by anglers. *CSO Team / Colby Simms Outdoors LLC*

This SI muskie battles it out boatside with CSO Team tournament champ Craig Fisher. *CSO Team / Colby Simms Outdoors LLC*

Lures that mimic schooling baitfish are top choices for top predators like muskies, stripers, and largemouth. *CSO Team / Colby Simms Outdoors LLC*

Pro angler and network TV personality Colby Simms hoists a powerful SI flathead catfish. *CSO Team / Colby Simms Outdoors LLC*

CSO Team's Walt Krause shows off a whopper Southern Illinois crappie he caught fishing with the author. *CSO Team / Colby Simms Outdoors LLC*

The author carefully works the figure eight on a following muskie during a cold and cloudy Southern Illinois day. *CSO Team / Colby Simms Outdoors LLC*

The magnificent colors of one of Southern Illinois's many beautiful sunfishes. *CSO Team / Colby Simms Outdoors LLC*

A huge rock bluff wall plunges into deep clear Southern Illinois waters where huge fish lurk. *CSO Team / Colby Simms Outdoors LLC*

CSO Pro Staff director Ray Simms caught this monster SI bass on a CST spinnerbait. *CSO Team / Colby Simms Outdoors LLC*

8

Premier Presentations

And we move right on with patterning. In this chapter we'll discuss all of the most effective tactics used to catch each of the various species of fish in Southern Illinois individually. Species by species we'll break it down. One by one we'll provide anglers the best ways to tempt each individual type of fish to bite, once the season and conditions have been taken into account and concentrations of active fish have been located.

Water conditions, weather conditions, and available prey top the list for consideration when developing effective patterns. The presentation is created with a set of fishing methods or techniques, deployed in the attempt to get fish to strike an offering. Some tools such as lures, baits, hooks, and other terminal tackle items make up a part of the presentation itself. Other tools like rods, reels, and lines allow us to properly present various offerings to the fish. Everything involved in the process is important, and no detail can be overlooked. Paying attention to the details can turn a good angler into a great angler. It can significantly increase numbers of fish caught over the course of weeks, months, and years. And this is especially important when it comes to the size of fish caught. Bigger, older fish are more difficult to coax.

When talking tactics, we decide which rods, reels, and lines are needed, which artificial lures or natural baits will work best, and which terminal tackle items are employed to seal the deal. We must also consider whether we're targeting fish with vertical or horizontal presentations, the speed at which to present our offerings, and the cadence of the retrieve. Basically, how we manipulate lures or baits with tackle in order to coax the strike from a fish. In most cases, we attempt to convince a predator that what we have at the end of our line is not only real but preferably injured. Or at least weaker than other prey to elicit a predatory response.

It is true that one day with perfect conditions could produce a hundred bass, while the very next might produce just ten bass all day long in poor conditions. But what is far more likely is a scenario that goes like this: On a prefront day with significant winds out of the west or the south, dark clouds, and an approaching storm (optimal conditions basically), many anglers on the water will catch 30, 40, or maybe even 50 bass. But at the same time a much smaller percentage of anglers, just a handful of anglers on the same water, might catch 90 or 100 bass, or even more. And with larger fish in the mix. Then after the storm has rolled through the region and the cold front has set in with clear skies, things change. Under flat calm conditions or light east winds, a small handful of anglers who work hard all day might land 10, 12, or 20 bass, with a few decent fish and maybe a big one or two. At the same time, the majority of anglers catch just a couple of small fish all day long. Most will not land any big fish, possibly even nothing at all. This is what we anglers refer to as the skunk, or getting skunked.

In most cases, the major differences between the groups of anglers, on both the good days and the bad, are those who work hard to develop a new pattern and do so with an open mind. And the open mind plays a huge role. Versatility and a willingness to learn and use new techniques, getting outside of one's comfort zone, play key roles in consistent success on the water. We all have our favorite lures, favorite methods, and favorite fishing styles. Chances are we're far better at using these than others. But good anglers will spend enough time practicing with the lures and methods that we don't like to fish. Then we can employ these with success when the fish refuse to bite on our old favorites.

Of course, it's always a good idea to start on the same spots with the same lures following a successful day on the water. But if this doesn't produce in fairly short order, and it often will not once conditions have changed, it's time to experiment and start working to develop a new pattern. Based on what was happening before the weather change, when conditions were good and the bite was hot, anglers will probably have to employ at least one if not both of the following methods. One is covering a lot of water, moving fast, and fishing many spots in different areas and various depth zones. Different places than the locations where the high activity occurred before. The second is to change up presentations. Rig up a lot of rods, constantly change between many different types of lures, and regularly vary speed and cadence.

We've previously discussed some of the top site-specific tactics for the more popular species at Southern Illinois's premier waters. Now we'll get into detail about many more fishing methods and techniques that have proven effective for the various fishes across the region. Southern Illinois tried-and-true tactics that you can take to the bank. Some of these are old favorites, both on our waters here and across the continent. Others are unique and highly specialized tactics that few anglers employ with regularity but produce big time, often when all else fails. Mastering these tactics and being able to quickly and efficiently switch between them on the water, make a big difference in nailing down precise patterns faster. Anglers can then put more fish and larger trophies in the net.

We'll break this down into subsections for the major groups of species, discussing the best patterns for each species individually. The first section on bass will be the longest, as we'll more precisely dissect tactical considerations. Remember, though, that many principals discussed in each species section can be applied to other predator fish. As we discuss various tactics, keep in mind to follow Illinois's up-to-date statewide fishing regulations, and also those site-specific regs that can vary with each different body of water. Let's dive in!

Black Bass

Around the beginning of each calendar year, a week or so after winter has officially started, it has already been cold enough for long enough to significantly drop Southern Illinois water temperatures. While most fish species can remain fairly active through portions of December, it's usually around the holidays that activity for all species really slows down and anglers must adopt specialized winter fishing methods. We've discussed Southern Illinois fish locations through the seasons. But the types of lures and tackle we select, and the fishing techniques and methods we employ, can be just as specific, or even more so.

Throughout the winter period, we'll be fishing slowly for most fish species, including America's favorite fish, the largemouth bass, and its close cousins, the smallmouth and spotted bass. Fishing tactics remain similar across the three species of black basses in this region. A top method for wintertime bass hunting is jerkbait fishing. Jerkbaits, also sometimes called minnow baits, are long slender plugs with diving lips on the front. A very old style of lure, jerkbaits have been catching fish for longer than lures have been commercially made and sold to the public. While most of the old standby versions were carved from wood, most of the newer styles today are hard plastic. Shallow and deep-diving versions work well in winter to take these hard fighting fish, depending on where they are holding, in sizes ranging from about 3-1/2 to 5 inches.

While floating versions of these lures work better later in the year when water temperatures are warmer, suspending models get the nod when it's cold. These weighted lures are designed not to float or sink when paused but to sit still in the water column. They hold their position, right in the face of lethargic, inactive fish that are not in the mood to chase something down. While a few jerkbait models will drop extremely slowly, most that are marketed as suspenders will either sit perfectly still or rise toward the surface at a very slow rate of speed. Water temperature plays a small role with fall rate too, since the density of water decreases as water temperature increases. Still, while perfect neutral buoyancy is optimal, an extremely slow-falling or slow-rising jerkbait often gets hammered just as well. Another option is using stick-on weights. These thin lead strips, with adhesive backing, are used to make floating jerkbaits suspend, and they can be adjusted to the perfect buoyancy. Make a long cast, and crank the lure down to its working depth. Give it a snap of the rod tip, and then let it sit.

Sometimes short pauses of just a couple of seconds will work. Other times extended pauses are best, sometimes painfully long pauses of 15 to 20 seconds, or even more. It can drive an angler a little batty to fish that slow. But in really cold water, very long pauses usually outproduce shorter ones. A few of my CSO pro staffers in Kentucky will "soak" a jerkbait for up to one minute at a time on Kentucky Lake and Lake Barkley. For a power fisherman like me, it's a nerve-racking tactic. But it's one that can reap tremendous rewards, for anybody who can develop the patience. I have to admit, I don't let a jerkbait sit for a full minute very often. But on Southern Illinois waters, I've nailed some big winter bass and other species by pausing a jerkbait for a good 30 seconds, or a little longer. This works exceptionally well in open water where you've marked suspended fish on sonar, or over and around boulders, brush piles, or standing timber.

Other top options for winter bass include jigs, which are paired with either pork chunk or scented soft plastic chunk or craw trailers. Fishing slowly is best, with either a drag-and-pause retrieve or short hops off bottom and pauses. Sometimes dead-sticking the bait for a long period of time works better. This is also effective with either Texas- or Carolina-rigged soft plastics fished alone. Crawfish imitators and freakbaits are top options. Safety-pin-style, long-arm spinnerbaits are a good choice. And slow-rolling a 3/4- to 1-ounce double or triple willow or hatchet-bladed version can take huge fish. We work them along the bottom. Bump the spinner into cover and then kill it, allowing it to fall to the lake floor.

A single-bladed short-arm safety-pin spinnerbait, on the other hand, is a great vertical jigging lure for suspended bass around baitfish schools in open water. I prefer the hatchet blade here. Other good options for vertical fishing are slender jigging spoons, and vibrating blade baits from 3/8 to 1 ounce. Adding additional spray-on scents to lures can increase wintertime bass catches. One last winter lure choice to mention here is the Alabama rig. Alabama rigs seem to really shine when it's cold, far more than any other time of year. I usually employ a standard 5-arm rig, with 4-inch paddletail-style soft plastic swimbaits.

As winter fades to spring and waters warm, a wider variety of options exist. We begin fishing faster, with moderate speed retrieves getting the nod most of the time. Throughout spring, various groups of fish will be in different stages of the spawn. Some fish will be in the prespawn, while others are spawning. And some have finished up and are in the postspawn stage. This occurs with all species of game fish, bass included. A variety of tricks work to catch bedding bass that are in the spawn. I have always recommended however, the anglers quickly photograph and carefully release any and all spawning bass that are taken off of beds, to help ensure future populations.

In the shallows, anglers can see the fanned-off depressions on the bottom of the lake. These are usually in protected areas, where bass use their tails to clean off silt and debris and create what is referred to as a nest. Smaller males arrive first, clean off an area, and fan out the bed. Then when the moon phase and water temperatures are right, the female will move in for a while. She'll eventually lay eggs that the male will fertilize. The female then moves back out of the spawning area. She'll often remain fairly shallow in heavy cover, or along the edge of the shallow break not far from deeper water. The male remains to protect the eggs, which can be a significant job. Anglers have a much shorter period of time to try to catch a big female bass off a bed. But they have plenty of time to target the male, and this is similar with many other species. In fact, many game fish spawn in similar ways. And many of these principals can be applied across the species spectrum, with differences in presentation.

One of the best tools developed to aid in bed fishing is the Power Pole. Using it to silently hold the boat within the perfect pitching range of beds has eliminated most of the difficulties of targeting spawning fish that are related to wind and boat positioning. Bass don't feed much during the actual spawn, as is the case with many game fish species. So, we have to play upon their instincts in order to get them to strike a lure. One trick is to make a mess. When something washes onto a bass bed, they will instinctively want to remove it. A male bass doesn't want anything in its bed before the female lays the eggs. A clean bottom helps increase the chances of successful reproduction. As such, females select mates that make a nice bed and keep it clean. By pitching something into the bed, the male will often remove it. Lures like soft plastic jerkbaits, craws, and stick worms are good options to pitch onto beds.

Unweighted lures can be used at times and are often preferred if possible. While bass might be able to see our boats, or see us targeting them from the shoreline, we still want to move slowly. Be quiet and stay as far away as possible while still being able to see the bed and the bass clearly, and make a good pitch. We don't want to spook the bass any more than is necessary so that they behave more naturally and go about their business. Making a quiet and subtle low-to-the-water pitch is best. Unweighted plastics don't make a big splash as they hit the water, so close to the fish. Sometimes, they'll gently pick up the bait to move it away, and an angler can set the hook and land the bass. Other times, though, bass swim up to the lure and blow it off the nest without ever touching it. They do this by opening and closing their mouths to forcefully push water that washes the item away. Either way, weight might be required.

It's always best to have multiple pitching rods rigged up and ready to go. A similar bait that's weighted might not be blown off the nest, requiring the bass to pick it up, provided it's not spooked too much from the cast. Small weights from 1/32 to 1/8 ounce can sometimes be enough. But we've jumped up to 1/4- or 3/8-ounce weights if needed. These heavier weights also help deal with wind or current issues that make it more difficult to keep a soft bait on a bass bed. I've had the best success with a sliding bullet sinker, usually pegged on a Texas rig. But jig heads and weighted offset hooks have produced well too. My longtime friend, champion bass tournament angler and CSO pro staffer Jim Lyle, on Lake of the Ozarks in Missouri, is the only angler I've ever known who prefers Carolina-rigged plastics for bedding bass.

I was a bit skeptical at first. But after being completely outfished one spring day, I switched to his rig. And we went almost fish for fish the rest of the trip. Of course, aquatic vegetation is not an issue on a reservoir like LOTO, with fluctuating water levels. On lakes with significant weeds and grasses, I don't use the Carolina rig in the shallows during spring. But it works on most lakes without lots of grass, and on rivers too. Carolina rigging usually works best when it's tossed beyond the bed and slowly pulled into it. While heavy weights are often the norm for Carolina rig fishing most of the year, at this time, smaller sinkers from 1/8 to 1/2 ounce are fine. Bullet weights will work, although the egg sinker is my preferred C-rig weight most of the time.

A second tactic is to attempt to trick the bass into actually eating a lure for food, by triggering its predation instincts. While bass aren't focused on eating, and most won't consume any food to speak of for some time while they are spawning, some can fall victim with the correct application of speed. Throughout the entire year, one method that we pull out of our bag of tricks when the bite is slow is to significantly speed up our presentation. When a lure rushes past an inactive fish at extremely high speeds, they may strike out of instinct. This goes along with that old cat-and-mouse concept. A cat will play with a mouse as long as the mouse is trying to escape. When the mouse plays dead, the cat often loses interest and may even stroll away. But when the mouse springs back up and speeds away, the cat immediately attacks. It usually runs full speed and jumps on the mouse, jaws open, often killing it immediately with a bite to the neck. Yes, I grew up on a farm with lots of cats. And no, you can't live on a farm without cats.

Anyway, it's the same concept with bass, bluegills, muskies, and most other game fish. Or even predatory rough fish. When a lure rushes past them extremely fast, they can strike without thinking. The key here is to select a lure that can be fished effectively at high speed. Top choices would include a lipless crankbait, a heavy spinnerbait with willow leaf or hatchet blades, or a heavy swim jig with a thin swimbait trailer. A bait-casting rod matched with a high-speed retrieve reel, with a gear ratio of close to 7:1 or higher, really makes this presentation and triggers some bedding bass to smash a fast-food meal.

Both of these tactics can shine for bedding bass. And sometimes one of the two is the only way to get spawners to bite. But we've saved the best for last. The optimal way to catch bedding bass is to trigger an aggressive attack on our offering. When a bass strikes something hard, out of aggression, we have a better shot at getting a solid hookset and driving the steel deep into their jaw, so that we can fight them to the boat and actually land them. The high-speed approach works well for this. But day in and day out, it triggers fewer actual bedding bass than placing something onto the bed. So how do we place something onto the bed and get the bass to aggressively attack it? By making the bass believe that it's a creature that will eat its young.

Turtles will eat fish eggs. And as such, small turtle-shaped soft plastic lures can take bass from beds at times. But this is not a huge threat to the bass. Bluegills, redears, and other sunfish eat both bass eggs and bass young. And you'll often see a bass chasing these smaller fish away from its nest during the spawn. This is why tubes often work so well to catch bedding bass. Tubes resemble fish well. All kinds of tubes can take bedding bass. But some tubes are better than others. The sunfishes are short and tall compared to most other game fish that are longer for their body size. Early in the spawn, when bass are hanging around the nests but not ready to drop and fertilize eggs, standard tubes can work very well. But when the males and females are close to actually doing the deed, and for some time after as the male protects the nest, fat tubes are great.

These thicker-bodied lures seem to resemble panfish better. The other important factor is the color pattern. Lots of colors can work well earlier on. But once the bass are really locked on the beds, color patterns that closely mimic bluegills and other sunfish in the area get the nod. Small and realistic panfish-style swimbaits can be effective too. But the way a tube glides into a nest can be magical. And while I've caught a lot of bass on various lures, at waters across the continent, these seem to shine here in Southern Illinois. There might be just one lure style that works as well as a tube for bedding bass. The lizard has been around for a long time and is a staple in the tackle boxes of all bass anglers. The lizard mimics a salamander, which will eat bass eggs. Bass often strike lizards hard, especially during the spawn. And this can be the hottest option for bedding fish at times.

Of course, anglers don't have to target bass on the beds to experience great springtime angling. As discussed, bass and other game fish spawn in waves. Fish move in and out of spawning areas throughout the season. Some anglers love bed fishing. It is exciting, especially in clear waters. But given the choice, I usually opt to target bass that are not spawning, for several reasons. While I do it all, I'm a power fisherman at heart. I just enjoy fishing fast, covering water and utilizing power tactics whenever I can. After so many years of full-time sport fishing for my living, it's been my experience that more big fish are taken with power tactics. And I would much rather catch fewer but bigger fish than the other way around. Still, bedding bass seem to be finicky more often than those in the prespawn or postspawn modes. And targeting fish that are not spawning is a simpler and more fluid prospect.

So, while some anglers are sitting on beds all day, my fishing partners and I are often backed just a little further out, targeting the rest. And it should be noted that most of the time there seem to be higher numbers of fish in other modes than those actually spawning. Faster fishing is the norm here. We stay on the trolling motor, covering water and targeting any and all shallow cover. Usually anywhere from the shoreline out to the first drop toward deeper water. Spinnerbaits with double Colorado or double Indiana blades, or those with either a double or triple set of hatchet blades are a top option in 1/4-to 1/2-ounce sizes. Swim jigs and bladed jigs matched with paddletail swimbait trailers are also top choices in similar sizes. Soft plastic jerkbaits around 3-1/2 to 5 inches take fish well now too.

Buoyant, shallow-diving crankbaits and floating short-lipped jerkbaits, from about 2-1/2 to 4-1/2 inches both take fish with this aggressive angling style. If the bite isn't fast and furious, sometimes we upsize or downsize. But most of the time a standard-sized bass lure puts lots of quality fish in the boat for us. When fishing is slower, Texas-rigged soft plastics can produce well. At this time, I usually prefer crayfish imitators and freakbaits from 3 to 4-1/2 inches. These are fished on a wide gap offset hook with a 1/16- to 1/4-ounce bullet sinker, sliding freely on the line.

As the waters continue to warm in late spring and the spawn is wrapping up, the topwater patterns gets rolling. This can be some of the most exciting bass fishing of the year, as everyone just loves a good surface bite. While bass often take topwater lures very well throughout most of the warm half of the year, it's primarily an early morning and late evening game. From late spring through early summer though, topwater fishing can be the hottest ticket going all day, and all night long. In fact, this is probably the only time of year when topwater fishing is regularly consistent right through the middle of the day, with even clear skies and bright sunshine. One great trick is to use a front runner. This is a small floating lure with a single belly treble hook and a line tine on both the nose and tail of the bait. Tie this onto the main line, then tie on a leader and attach your floating topwater lure, like a chugger or propbait, to the end. It can look like a predator chasing prey, or like schooling fish swimming together. But it often results in double hookups on bass, and other species too.

On most Southern Illinois waters, buzzbaits, typically from 1/4 to 1/2 ounce in weight, are a top choice at this time. These allow anglers to move fast and cover a lot of water in search of aggressive fish. Tail-spinning propbaits, from about 3-1/2 to 4-1/2 inches, are great at this time too. Poppers and other chugging-style lures, from around 2-1/2 to 4 inches, as well as 3- to 4-1/2-inch walking-style cigar plugs also work well now. Wake baits, which are a kind of hybrid cross between a topwater lure and a shallow diving crankbait, are fished effectively on top during this period in similar sizes.

Other hot lure choices for bass now include soft jerkbaits from 3 to 5-1/2 inches, jerk worms, trick worms, and similar straight-tail soft plastics from 4 to 7 inches, as well as various standard and hybrid swimbaits from 3 to 5 inches in length. We're also still catching some bass on shallow diving crankbaits, as well as double- and triple-blade long-arm safety-pin spinnerbaits in 1/4- to 1/2-ounce sizes. But an underfished lure for bass can be deadly much of the time now too, the inline spinner. Traditional, single-blade inline spinners are very popular in small sizes for trout, bluegills, and all sunfish species. And the large versions of the lures have always been a staple in the tackle boxes of anglers targeting muskies, as well as their cousins, the northern pike and pickerel. But medium-sized inlines get very little attention in the bass world, even though these lures take largemouth, smallmouth, and spotted bass well at times. The late spring and early summertime period can be especially good for bass with inline spinners, particularly those ranging in size from 1/4 to 3/4 ounce.

Fishing slows down a bit in middle summer, but tactics get simpler. Basically, it comes down to shallow water or deep-water fishing. Early morning and late evening are the best times to target the shallows. Top choices of fast-moving lures include swimjigs with paddletail swimbaits, and double- or triple-blade long-arm spinnerbaits in 3/8- to 1/2-ounce sizes, as well as 3/8- to 3/4-ounce shallow-diving cranks. When fishing is slower, 1/4- to 1/2-ounce flippin-and-pitchin-style jigs with bulky skirts and big plastic craw trailers can work. Oversized worms can be deadly in summer too.

My good friend Charlie Ewen has always had great confidence with big plastic worms for largemouth bass. Fishing tournaments on Lake of the Ozarks with his dad Tony for many years, it's no wonder, since worms have always been a top bait at Missouri's biggest reservoir. While I don't fish worms that often, when I have employed them for Southern Illinois bass in the past, they have mostly been small or medium-sized offerings. After Charlie bested me on quality largemouth two days in a row on Kinkaid Lake, and then on a pair of the lakes at Pyramid State Park, I gave in and started throwing big worms for bass. That pattern produced well much of the summer, and it has been effective on various waters off and on for years since. In fact, when worms are the ticket for big bass, but they're not crushing medium-sized offerings, one would usually think that going to a smaller size would produce better. Often however, the opposite is true. And they'll hammer a large or even oversized worm on a big hook.

Curly-tail, ribbontail, and even straight-tail versions in 8- to 12-inch sizes, produce big bass when it's hot, on 1/8- to 1/2-ounce bullet-weighted Texas rigs in the shallows. Choose a big hook like a 6/0 wide gap. Topwater fishing remains a great option all summer long, especially early and late in the day. Buzzbaits are still a top choice. But this is when the frog comes into its own. Plastic frogs are dynamite for big summer bass. They can be fished without snagging, over even the thickest matted vegetation where hot-water bucketmouths often lurk.

For deep-structure fishing in summer, long-billed deep-diving crankbaits from 1/2- to 1-1/2-ounce sizes are highly effective. Slow roll big, long-arm spinnerbaits from 3/4 to 1-1/2 ounces, with double or triple willow leaf or hatchet blades, or drag and hop heavy jigs with craw trailers, or those 8- to 12-inch worms with 3/4- to 1-ounce sinkers. While Texas rigs work for big worms and other soft plastics, summer is when the Carolina rig really shines. By using a leader of 10 to 20 inches to separate the soft bait from the sinker, we allow the lure to float and glide more freely in the water. This seems especially attractive to bass and other predators in the heat of summer.

A special trick that's seldom used, is to replace the big sliding sinker with a heavy jig of about 1 to 1-1/2 ounces. The leader can be tied to the back of the hook or off the line tie. And now you have two different lure options for bass. Sometimes they take the jig, other times the soft bait trailing behind it. Sometimes we catch two fish at a time, with one on each lure! Trimming the jig skirt and selecting a smaller trailer reduces resistance in the water, to get the rig down faster. A big straight- or ribbontail worm or a large soft jerkbait can appear as a predator fish, chasing a crayfish scurrying along the bottom. Night fishing is a great option in the heat of summer, and all of the shallow and deep fishing tactics mentioned here will produce well at night too.

As late summer begins to fade into fall, we settle into a long period of shallow-water bass fishing. Medium- and high-speed retrieves get the nod on most days. Activity levels are often high, as bass make their biggest weight gains of the year. We don't downsize very often now. Both standard and oversized lures work well for bass from late summer all the way through late fall. Buzzbaits from 3/8 to 1 ounce, buzz frogs from about 2-1/2 to 5 inches, and propbaits from 3-1/2 to 6 inches are all excellent topwater options. If I had to choose just one lure that would be my favorite to catch bass on, the most fun anyway, it would have to be the buzzbait. A simple yet deadly choice, watching it gurgle and sputter its way across the surface just keeps you on point the whole time. And the way that a big bass savages this lure is nothing short of heart-stopping. Yes, I just love buzzbaits for bass!

Swim jigs and bladed jigs from 3/8 to 1 ounce, matched with 4- to 6-inch paddletail or split-tail swimbait trailers are hot now too. Lipless crankbaits and shallow diving cranks from about 2-1/2 to 5 inches and floating jerkbaits from 4 to 6 inches produce too. Long-arm safety-pin spinnerbaits are the favorite lure of both Dad and Amber, and are one of the best lure choices at this time. We usually select them in sizes ranging from 3/8 to 1-1/2 ounces, with 2, 3, or 4 hatchet or willow leaf blades, and often add 3- to 5-inch paddletail swimbait trailers.

Speaking of swimbaits, fall is the best time for swimbait bass fishing in Southern Illinois. Different than the soft plastic swimbaits, which are primarily used in conjunction with either straight jigheads, wide-gap weighted hooks, or the classic skirted weedless swim jigs, classic swimbaits have molded-in frames, weights, and either single hooks protruding from the bait or hook hangers that trebles are attached to. This style is categorized as one lure. Whereas the jig and soft swimbait is actually two different lures paired together. And the action can vary quite a bit.

Classic swimbaits get most of their action from the tail. Hybrid swimbaits, on the other hand, get most of their action from a short, nearly vertical lip molded into the bait frame. Both have a highly realistic shape and appearance. Some solid hard-plastic-lipped lures are marketed as swimbaits but more closely resemble crankbaits or minnow baits. But there is one other unique hybrid swimbait to mention. These lures have hard plastic heads, with a soft plastic body and tail, or a hard plastic head and body with a soft plastic tail but no lip. They have segmented bodies and swim in a kind of serpentine action underwater. The most common sizes for fall bass here range from 4 to 6 inches.

Faster-moving horizontal presentations are usually the best place to start in fall. But when these bites slow, a vertical approach is sometimes needed. One of our original CSO pro staff members, Andrew Veach, is one of the best muskie anglers I've fished with. Dad and I first met Andrew and his dad during our inaugural season on the Illinois Muskie Tournament Trail, competing against them. With our shared passions for fishing and other sports, Andrew and I became good friends. With the fishing business growing, Andrew joined our team and began guiding fishing charter trips for us, as well as testing our CST tackle products. I had Andrew appear with Dad and me on a couple of *Adventure Sports Outdoors* TV shows on Southern Illinois bass and muskie fishing. We traveled, fished tournaments together, and went on a saltwater trip to Venice, Louisiana, that turned out to be the best week of redfish angling of my life. So, we've had quite a few adventures.

While muskie fishing introduced us, we've actually fished more for bass together than anything else. And we've both learned techniques from each other. While I've almost always been a power fisherman, moving fast with horizontal presentations as often as I could, it was Andrew that got me back into jig fishing. I had fished vertically with jigs a lot when I was younger. But, with the exception of swim jigs and a horizontal presentation, I hadn't done a tremendous amount of jig fishing in a number of years. Andrew and I fished a June bass tournament on Kentucky Lake. With little wind, blistering heat, and bright sunshine in the forecast, we were expecting hot weather, not hot fishing. But the one thing we had on our side was the fact that they were pulling a lot of water through the dam most days. And this was triggering fast action when the flow was high, but only on specific mid-depth ledges near deep water drops.

In practice, we were able to find concentrations of big largemouth bass on a number of ledges that got fired up when the water was flowing. Hopping and dragging heavy weedless football jigs with plastic craw trailers was a hot ticket that helped us score a quality sack in the event. But jig fishing is one of Andrew's favorite tactics for bass on all waters, including those in Southern Illinois. Sometimes it takes something like this tournament to spark renewed interest. And since then, I've been using jigs with a bottom presentation to score big SI bass all year round. This is always a top choice when a vertical presentation is productive throughout fall. And it carries over well to produce big bass in wintertime too.

Speaking of fall and winter vertical jig fishing. Soft plastics usually rule the day when it comes to jig trailers. But don't forget the old tried-and-true pork chunk, especially in the cold-water season. I was never a big fan of pork trailers. They're kind of messy. And I always figured that heavily scented soft plastics worked just as well. That was until the day that Chris Shannon and I were on a tough bite for Southern Illinois bass and he turned around by switching to pork chunk and eel trailers for his jigs and spinnerbaits. A longtime friend, hunting partner, and CSO pro staffer, Chris is a big fan of adding various pork trailers to entice trophy fish. Fishing them slowly has produced some whopper largemouth when they wouldn't touch the soft plastics. And they can be a game changer from late fall, through winter, and into early spring, especially when action is slow.

Day in and day out, we're using mostly bait-casting rods and reels for bass fishing. For those who don't like this kind of equipment, you should really consider spending the time it takes to master it. Because once you make the switch, you won't go back. Bait-casting equipment allows for better and more precise presentations. It helps anglers to place lures exactly right in the spot where they need to be, and with less noise. Sometimes laying a jig, spinnerbait, or Texas-rigged worm right against a stump will elicit a strike, while landing a few inches away means nothing. For most standard-sized offerings, we select rods from 6 feet 4 inches to 7 feet 2 inches in length. These are mostly rated for anywhere from 3/16- to 3/4-ounce lures, in medium light to medium heavy power ratings. For bass it's mostly fast to extra-fast actions. Although we will use moderate action cranking rods, or a moderate fast for hard surface plugs.

We do utilize some heavy power flippin sticks at times, usually 7-1/2 feet with an extra-fast or ultra-fast action, rated for 3/8- to 2-ounce lures. We keep a few spinning outfits handy as well, mostly for finesse fishing with small soft plastics or other downsized lures. These can be anywhere from 5 feet 6 inches to 7 feet, from ultralight to medium-light power ratings, and moderate to fast actions, for lures from 1/32 to about 1/2 ounce in weight.

For spinning reels, we look for higher-speed versions with larger spools. But bait-casting reels are used for probably 95 percent of our bass fishing. And we really love the new compact ultra-low-profile designs. They make for a much more comfortable day of fishing. They also aid in accurate casting, retrieve cadence, and hook setting by allowing an angler to better wrap the hand firmly around the side of the reel. Dad and I both prefer a 5.6:1 gear ratio for mid-depth and deep-diving crankbaits, a 7.1:1 reel for topwater lures, lipless crankbaits, hatchet or willow leaf blade spinnerbaits and swim jigs, and 6.4:1 for most other presentations. For most bass fishing, we choose 8- to 20-pound monofilament or fluorocarbon lines, or braided lines testing 20 to 50.

Sometimes, slower finesse-style fishing methods are required to get bass to bite when the action is slow. However, bass are typically an aggressive species that can often be taken with high-speed power-fishing presentations. Dad and I have both caught more and bigger fish, and we've won more tournaments fishing fast, with power tactics, than any other. When things are tough, speed gets your offerings in front of more fish and gives the bass less time to think, triggering reactionary responses from some. While slow finesse methods must be in the bag of tricks for all anglers, for bass it's usually best to start with a fast approach and power-fishing tactics, only slowing down if need be.

Crappie

Wintertime crappie fishing usually occurs in deep water. On warmer days, particularly those with bright sunshine, the shallows will warm a bit, and some fish of various species may move up into the shallower water to feed. This can especially be the case with multiple days of nice weather conditions. In this instance, crappie can be taken from shallower water. But most of the time it's going to be a deep-water bite. Winter crappie fishing is a fairly specific prospect, and specs are primarily caught with a couple of presentations, one live bait and one artificial. Crappies love minnows, all year round in all conditions. Crappies will take other live baits, like mealworms, wax worms, red wigglers, crickets, leeches, and even small crayfish. But day in and day out, anglers will catch so many more crappies with minnows that these other live baits are almost never needed. Both shiners and fatheads seem to work equally well. A medium minnow is usually best. But we sometimes drop down to a small in winter, when fish are especially sluggish.

The second top tactic for winter crappies is jig fishing. Jigs are a consistent artificial lure option all year long, and even through the coldest part of the winter months in sizes ranging from 1/64 to 5/16 ounce, with body sizes from around 1-1/2 to 3-1/2 inches, although the typical sizes for most applications would be a 1/32- to 1/8-ounce jig with a 1.75- to 2-1/2-inch body. A wide variety of lead heads can be effective for these beautiful and tasty panfish. The classic marabou jigs are great in winter. The tiniest manipulation of the rod tip can cause the soft hairlike material to dance seductively, to trip the trigger of a crappie's instincts and cause it to strike. And other various hair jigs can produce too.

Plain jigheads matched with specific soft plastic lures can work great. Hollow-body and solid-body tubes are a top choice for still fishing or using a slow retrieve, tactics that are often required for big winter catches. The tentacles wiggle with little movement. Small worms can be effective, especially straight-tail versions. The other top choice of soft bait in winter would be one of the tiny shad-type lures. These little single-tail swimbaits, with their skinny needle-pointed tips, will quiver in the water with the slightest movement. It's a subtle presentation that is ideal for sluggish crappies in very cold water.

Some days jigs will produce better. Other days it will be live baits. And sometimes the two bait choices will work equally well. So, it's best to experiment with both until it becomes clear that one will outdo the other. The second biggest choice you'll make at this time is rigging. Floats can be important. Fishing a jig or a minnow under a float allows anglers to cast the offering next to the standing timber, laydown trees, stumps, brush piles, boulders, broken rocks, bridge pilings, and docks, where wintertime crappies will often hold. And most importantly, the float allows the angler to keep the offering in the strike zone for an extended period of time, at any depth, with little movement. Oftentimes, less is more for winter crappies, and it almost becomes like a soaking or dead-sticking technique. This can work great for targeting mid-depth cover, or shallow cover when the crappies move up. Fixed floats can be used for shallow-water fishing, while slip floats are required for moderate depths or deep water. These pair well with both jigs and minnows.

The other option with jigs is to cast and retrieve if fishing shallow or moderate-depth water. This can be productive at times, especially when crappies are a bit more active. But the key is usually to fish slowly, barely moving the jig along. Fishing vertically without a float also produces at times, right over top of the fish in deeper water. Or by using a long crappie pole to reach cover, and getting the lure out away from the boat when fish are shallower and you don't want to spook them. This is an even more important consideration in clear waters or those with higher fishing pressure. With this method, often called dipping, anglers can really load the boat quickly when stained-water crappies are bunched up in big numbers tight to cover.

A highly effective and underutilized technique for crappies, anytime of year when they are found in moderate depths or deep water, is one that didn't originate with crappie anglers. Drop shotting, or drop-shot fishing, is a technique borrowed from the bass fishing world. Bass anglers developed the drop-shot tactic as a finesse presentation to take finicky fish on small, subtle soft plastic lures, fished just above bottom in deeper waters. Usually on clearer water lakes.

We first experimented with drop shotting for bass on the deep clear waters of Missouri's Table Rock Lake, but noticed that it produced huge bluegills too. The technique works equally well for crappies. And after devoting serious time with it in Southern Illinois, we realized that it might be better for crappie fishing than for any other species. Drop shotting has become one of our favorite tactics for deep specs in winter and summer. We primarily use this rig with live minnows in winter. Small soft plastics can work well too, especially the tiny shad-type lures. The single skinny tail does exceptionally well for getting the drop on them.

Tie a hook onto the main line, and leave a long tag end. Attach a small bell sinker, a big split shot, or better yet, a specialized tungsten drop-shot weight to the bottom of the line. The rig is dropped vertically to the bottom, and the lure or bait sits above bottom where it gets easily noticed. Upping the odds with this tactic, anglers can add a second hook and fish two different depths zones. I like to place one bait just 4 to 8 inches or so above bottom, and the second a couple of feet up the line from that. This also allows more experimentation.

You can use both small and medium minnows, or a shiner and fathead at the same time. You can use two different size plastics in different color patterns, or go with one plastic and one live bait, until they show a preference for one kind over the other. This rig often produces two fish at one time. One last winter option to mention can be deadly in the right circumstances: suspending jerkbait fishing for crappies is best in winter. We utilize this technique exactly the same way for crappies as we do for bass, except we select mostly the smaller models of lures from about 2 to 4 inches, as well as lighter lines and tackle.

For many people, even those who don't fish that much throughout the rest of the year, springtime means two things, mushroom hunting and crappie fishing. Crappie fishing is almost always better in spring than in other seasons. And frying up a mess of crappie fillets with morel mushrooms is a meal fit for a king. It just so happens that morel magically pop up from the ground during a time when lots of big crappies are being caught with regularity. So often the most difficult question is, how much time do we take off the water to get enough mushrooms to toss in the pan with the specs? Wild turkey makes a great compliment here too, but we won't complicate things any further. Except to say that for the outdoor sports enthusiast, there really is never enough time in the day, as the old saying goes. Maybe they should have said season, instead.

Springtime crappies are more aggressive in the warming waters. These fish abandon the deep and flood into the shallows, like spring rains cause streams to bust out of their banks and flood into the woods.

On the smaller number of days when the crappie bite is slow, live minnows fished under fixed floats usually still take enough fish. But we mostly prefer to stay on the move at this time, covering lots of water and hitting as much shallow cover as we can with artificial lures. It's faster and more efficient. The same marabou and hair jigs that produce in winter will take fish now too, particularly from 1/32 to 1/8 ounce. And the underspin jigs become more important as well. The same is true of soft plastics. However, many other plastics also begin working well now.

Tubes and small single-tail shad-type swimbaits are still top options. But double-tail swimbaits, single and double curly-tail grubs, reapers, flat-tail minnows, plastic worms, bug and crayfish imitators, and just about anything else you can think of designed for crappies will produce fish at times in the warming waters. And, lure size can jump a bit, with 2- to 3¾-inch plastics effective, depending on the type. Skinny worms for instance, are used in longer lengths than thick body baitfish and crayfish imitators, or the fat-style tubes that have more bulk. Faster and more aggressive lures and retrieves can produce more fish as waters warm through the spring season. In spring we begin casting and retrieving small spinnerbaits, bladed jigs, and jig spinners from 1/32 to 1/8 ounce, as well as small floating crankbaits and jerkbaits from 1-1/2 to 3-1/2 inches.

My cousin and CSO pro staffer Craig Fisher turned me onto vibrating blade baits for big crappies in Southern Illinois many years ago. He often preferred blade baits for white bass and panfish when we fished the Mississippi River, and especially the small feeder rivers and creeks flowing into it. Craig is a Southern Illinois bass tournament champion and big bass award winner. But he's equally adept at catching its other species, panfish included. Even though few panfish anglers utilize this category of lure for crappies, Craig's favorite blade baits become a hot ticket for specs as waters warm. After I netted one of the biggest crappies I'd ever seen to that point in my life, which Craig caught with a blade bait on a Southern Illinois lake one warm spring day, I decided to devote some time to them. I began catching a lot of big white and black specs on them and still enjoy that success with them to this day.

Crappies begin transitioning in early summer. At different times, we mostly find them in one of two places: shallow weeds and wood cover at the edges of flats near breaks to deeper water, or suspending in open water at moderate depths over deeper structures. For the shallow fish, casting bigger crappie jigs and spinnerbaits from 1/16 up to 1/4 ounce and 2 to 3-1/2 inches is a hot tactic. For the suspending fish, casting and trolling floating crankbaits and deeper-diving jerkbaits is a great and underused method. In this instance, we troll mostly 1/8- to 3/8-ounce moderate-depth-running 1-1/2- to 3-inch baits, from about 6 to 10 feet deep. But we may step all the way up to 1/2- or even 3/4-ounce 2-1/2- to 4-inch lures, trolled anywhere from about 8 to 16 feet deep.

As summer progresses, some of these suspending fish move slightly deeper while continuing to hold over deep structures in open water. They are susceptible to the same casting and trolling tactics with diving crankbaits and jerkbaits. Some fish will hold right on the structures at similar depths and can also be taken on deep cranks and jerks. In addition, summer crappies in deep water can be caught on 1/16- to 3/8-ounce jigs, vibrating blade baits, and jigging spoons from 1/8- all the way up to 3/4-ounce sizes. They also eat smaller-sized heavy spin-nerbaits, like the hidden weight versions, from 1/4 to 3/4 ounce in size with a smaller profile and appearance. And of course, live medium or large fathead or shiner minnows produce on jigs and drop-shot rigs, as well as split-shot rigs or under slip floats. Some fish will remain in shallow weeds all throughout summer. We take these most commonly on 1/32- to 1/8-ounce traditional jigs and underspin jigs, hair or plas-tic, and 1/16- to 1/4-ounce spinnerbaits fished over the grass, or on minnows fished under fixed float rigs at the weed edges, right next to the drop-off.

While fall does cause a general migration of bass and most species into the shallows, panfish and catfish are the primary species that can regularly be caught in deep water throughout the entire fall period on most waters. Granted, many of these fish remain shallow, and in some cases, extremely shallow throughout this season. But there will usually be some fish down deep, and sometimes they are more catchable. *Deep* being relative of course. The deep early fall crappies will still usually hang at or close to the thermocline. We used to catch most of our fall crappies from shallower water. But the bizarre weather patterns and changing conditions of our modern world have us changing tactics when we need to. And this bite seems to shift deeper more often as years pass. We still always start our hunt shallow, but we progress deeper if we're not catching enough fish. As a result, just about any tactic that we used for crappies in winter, spring, or summer can be effective for fall specs. And we have to switch between them until it becomes clear which is best.

We primarily utilize spinning equipment for crappie fishing. Most quality open-face spinning reels with spools that hold adequate amounts of line are fine. We do use some closed-face spin-casting reels for a few applications. Especially the underspin trigger reels, which we only use on long poles for dipping. Speaking of dipping, we use rods anywhere from 10 to 15 feet for this tactic. And for anyone that has spent much time fishing this way, they realize the value of poles that are light in weight. The majority of the time, it's classic ultralight- to light-power spinning rods from 5 feet 6 inches to 7 feet in length, with moderate fast actions, paired with 4- to 6-pound test monofilament line. We might drop down to 2-pound mono or a 4-pound fluorocarbon for spooky fish in very clear or highly pressured waters. And we sometimes jump up to 8-pound mono or 12-pound braid for lifting big stained-water specs with dipping poles.

Crappies are also called papermouths for a reason. It's easy to tear a crappie's mouth and pull the hook right out with a hard hookset, especially if the rod has too much backbone. Soft easy hooksets, or even just a lifting maneuver, are the norm with these panfish. This can be extremely difficult for a bass angler to do, or anyone that fishes for muskies or stripers or any big game on a regular basis. I was guiding a multispecies charter trip one day with a couple of out-of-state anglers that wanted to fish bass and crappie. The bass bite had recently been red hot, with most of the action occurring in shallower water in the morning hours. Crappie fishing had been a little slower, and most of the specs we had been catching that week had been coming in slightly deeper water in the afternoon and evening. After spending the entire morning slamming large 4/0 and 5/0 single spinnerbait hooks into the jaws of lots of big largemouth bass, with bone-jarring hooksets, switching over to crappie fishing was like climbing into a Pinto after having just raced a Corvette around the track.

Luckily, we were on plenty of crappies right off the bat. I say *luckily* because both of my clients missed fish after fish for quite some time. They were mostly bass anglers and hadn't done a lot of crappie fishing. But even if they had, after the kind of morning we'd experienced, almost anyone would have difficulty getting used to the switch. I explained the techniques to them and showed them what to do. Still, after 15 or 20 minutes, most of the strikes had resulted in lost fish. And they were talking about going back to the bass. I suggested we give it a few more minutes. I began saying in a low tone, "Easy guys, nice and easy." I repeated this every few minutes. And after less than half an hour, they were landing most of the crappies that struck their lures. We all limited out before the day was done.

One thing to remember with crappies too is that they often bunch up tight to cover. Oftentimes, we use faster artificial lures to cover water and find fish, and then slow down and catch more that won't hit a faster-moving offering. After we catch a couple of crappies from one piece of cover or one small area, we'll usually stop there for a few minutes and more thoroughly work the spot with the artificial lures. Then we might sit for a while and fish it hard with live bait, especially if one or more of the specs we caught on lures was a big crappie. This occurred one beautiful spring day, when I was doing a *MidWest Outdoors* TV show in Southern Illinois with my friend and cohost Matt Schultz. We had wrapped up a bass shoot that morning and then went after the crappies in the afternoon. After finding some specs while fishing faster with lures, we slowed down and switched to live minnows and caught a bunch more whites and blacks. We landed some big crappies, as well as some nice bonus sunfish. It's a good one-two punch for panfish.

Muskie

The run-and-gun is a high-speed fishing tactic that involves not just fast fishing but fast boating as well. Anglers typically choose stout baitcasting equipment and horizontal lures that work well at high speed, like spinnerbaits, buzzbaits, crankbaits, and swim jigs. They stay on the trolling motor and cover water on each spot efficiently, while making as many casts as possible. But they also change location frequently, firing up the outboard and running at top speed from area to area and spot to spot. This is done in an attempt to determine which parts of a large lake or river might be holding more active fish. And because this is the quickest way to figure out new waters, or those that an angler hasn't been on in a while, this is usually the preferred method for tournament anglers to begin their competition or begin a practice session before an event. This works great for bass, stripers, and other species, but might be more important to the muskie angler than anyone else.

Anglers use a lot of different kinds of boats for various styles of fishing. As the name suggests, bass boats, primarily of fiberglass construction, are specifically designed for bass fishing. Walleye boats, fiberglass and aluminum styles about equal in popularity here, are also their own category of fishing boat, best for anglers who mostly target walleyes. While not named for it, aluminum johnboats are often considered best for most catfish angling situations. And the center-console, inshore saltwater boats are the preferred vessels of most diehard striper anglers. All of these cross over and can work for different species. But when choosing a fishing vessel, it's best for anglers to first consider the species they target most frequently, and the way they prefer to fish. Most die-hard northern muskie anglers choose walleye boats for this sport. The bass boat has always been the ride of choice for Dad and myself. It's also a top choice for muskie anglers across the southern range. Yes, we do target all species, and we do fish for bass as much as we target muskies. But for us, the bass boat offers a ton of advantages for chasing the toothy tough guys. And many of our muskie fishing pro staffers choose bass boats as well.

For speed, the bass boat cannot be beat. As far as freshwater fishing boats go, bass boats are hands down the fastest on the water. They offer a significant advantage in most tournament situations, but especially with the run-and-gun. On large waters, anglers with bigger and faster boats can cover more water, hit more spots to narrow down patterns faster, and spend more of their day fishing instead of driving. This is just as important in or out of competition. The huge casting decks and low-to-the-water fishing platform makes working muskie baits easier for us. These also help facilitate a better response to boatside strikes and allow for a deeper and more effective execution of the figure-eight maneuver, which is unique to muskie fishing and critical to its success. With long rods and big nets, the bass boat also makes it easier for us to land and release these large, hard-fighting predators.

As a father-and-son team, Dad and I won the very first tournament that either of us had ever competed in. It was the opener event in the inaugural season of the Illinois Muskie Tournament Trail, which was on Southern Illinois's Kinkaid Lake. We racked up additional points through the season, and also went on to win the IMTT State Championship later that fall, up on Lake Shelbyville in Central Illinois. The run-and-gun tactic was critical to both of those wins. It's also helped us score other victories and high finishes in competition for muskies, bass, and billfish over the years. Speed and efficiency cannot be overstated when it comes to fishing success across the species spectrum. But the run-and-gun is most critical for targeting muskies. This is because there are far fewer muskies per acre of water than just about any species, and catch rates are low compared to just about everything else. Speed and efficiency help us get our lures in front of more of these rare toothy fish.

Muskies behave differently down here in the southern range than they do up in the northern United States and southern Canada. There are many similarities, but many differences as well. Most of the waters up north are perch-, whitefish-, or cisco-based systems when it comes to forage. Our waters are either shad or panfish based, and of course much warmer than in the north. Large-sized muskie lures are required for consistent action from adult fish throughout most of the season in a lot of waters of the north. And oversized lures, downright giant offerings, are becoming more productive and popular up there. Large and even giant lures do work in Southern Illinois. But we are fortunate in that during most of the year here, small and medium-sized muskie lures are just as effective, if not more so, for consistent action than the big stuff. And this is the case with fish of all sizes. Many of the biggest muskies taken here come on either standard or downsized muskie baits. Winter is the primary exception though.

In early winter and sometimes into the middle of the season, many muskies will spend considerable time in deeper water, and slower fishing is the norm. Some skis hold tight to structure and cover. But at least as many can be found suspending in open water zones over very deep water, out in the middle of nowhere. Vertical jigging can produce muskies during limited portions of the year, especially when fish are suspending. While casting is usually the most effective method to target muskies during most of the year, trolling often reigns supreme in the first part of winter and can continue into the middle portion of the season. Large-sized crankbaits and minnow baits from about 6-1/2 to 10 inches and 1-1/2 to 6 ounces are a top trolling option at this time. But they can be cast too. Giant cranks however, stretching the tape from 11 to 14 inches or more, with some maxing out at over half a pound in weight, have produced some wintertime whoppers for trollers.

Long-arm, 2-1/2- to 5-ounce safety-pin-style spinnerbaits with 3 to 5 willow leaf or hatchet blades are another hot option in winter, for trolling or casting. Conversely, a shorter-arm spinnerbait, with a single large hatchet or Indiana blade, excels for vertical jigging in the same weights. Big metal spoons from 1 to 4 ounces can also be jigged right now. Jerkbaits have a different meaning in the muskie fishing world than they do for bass and walleye sports. Classic muskie jerkbaits do not have diving lips like bass jerkbaits, or jerkbaits used for walleyes or crappies. What muskie and striper anglers typically refer to as minnow baits are referred to as jerkbaits by all bass anglers and many walleye and crappie anglers. Both the rise-and-dive-style and the glider-style muskie jerkbaits can take fish at this time, especially the weighted models designed to run deeper.

Minnow baits are productive in winter too. Suspending models or slow sinkers usually work best. But floating plugs with some added weight, to slow their speed of rise, take fish well. Rubberbaits, often called pullbaits, are a kind of hybrid lure. These giant soft plastic lures are made with a much thicker and more durable material than soft plastics for bass and other species. Many of them resemble swimbaits, with a variety of different kinds of tails. Big rubberbaits come in curly, ribbon, paddle, and forked-tail designs. You have single-tail models, double-tail models, even some with 4 tails or more. But these lures are molded onto a weighted frame with hook holders for big trebles. When selecting options in this type of lure category, during most of the year in Southern Illinois, we prefer large swim jigs paired with more traditional soft plastic swimbaits. But in winter, rubberbaits often produce well.

Muskie fishing can heat up fast in late winter and early spring, as these fish are quicker than most to increase activity with lengthening days, and even just slight bumps in water temperature. And especially in recent years, with warming trends and milder winters, activity can really spike in the middle of winter too. Muskies move into the shallows and start feeding heavier just as soon as they can, and presentation changes. Moving in tighter, we're trolling less and casting more as well as increasing presentation speed. We're tossing the same multiblade spinnerbaits as before, but in smaller sizes around 1 to 2-1/2 ounces. Cranking and twitching small floating minnow baits from 4-1/2 to 7 inches works well, usually with a fast and aggressive retrieve style. Lipless crankbaits come into their own now too. These tight wiggling rattlebaits work best for muskies from the middle of winter through early spring in sizes ranging from 3 to 5 inches and 3/4 to 2 ounces.

The lipless crankbait bite continues on into mid-spring but then dies off quickly. Topwater lures start to become effective at this point, with buzzbaits and propbaits working the best in sizes from 4 to 7 inches. But this is a great time to throw a chugger or a wakebait back at a fish that missed the faster-moving lure on an initial strike. A 6- to 9-inch soft jerk can work as a good pitch-back bait too. The smaller spinnerbait and minnow-bait bites continue on through mid-spring and late spring, and remain top options. The muskie swim jig bite turns on now as well. Long before swim jigs became popular with a significant percentage of muskie anglers, Dad and I were reaping their rewards. But we can't take credit for it. Earlier in my career, I traveled into the far north with Mark Davis for magazine work, and a once-in-a-lifetime experience. Luckily for me, this became more than a once-in-a-lifetime experience, in a place so remote that it takes two days of travel to get there.

Former National Guard Team fishing pro Michael Murphy and legendary outdoor sports journalist and magazine editor Soc Clay accompanied us into the land of the great ice bear to do battle with creatures of myth and legend, the giant Canadian pike of northern Manitoba. Not far from Nunavut, near the edge of where the boreal forest meets arctic tundra, we fished the famous North Knife Lake. It's 100,000 acres of virgin water, where many fish have likely never seen a lure before. We caught huge lake trout and chased whitefish and arctic grayling. But it was the water wolf that we were after, on one of the planet's best fisheries. This midsummer trip had us fishing some shallow rock, but primarily the heavy weed cover that dominates prime pike habitat at that time of year.

We caught mind-boggling numbers of big pike on a variety of presentations, including spinnerbaits, inline spinners, spoons, buzzbaits, and lipless cranks. But there were many areas that were just too thick to fish, with anything other than a swim jig anyway. Mark had brought a box full of big, weedless skirted jigs from about 1/2 to 2 ounces, and bags full of 5- to 7-inch paddletail-shad-style swimbaits that we attached to the back of the jigs. Unlike standard molded swimbaits with treble hooks, the swim jigs with big single upturned hooks and heavy fiber weed guards proved just the ticket to get over and through the slop.

The fishing was so incredible that Mark and I went back up to North Knife again the very next season with Dan Keith and Josh Raglin to film a national network TV show for *MidWest Outdoors*. Although the second trip was early in the season, just after ice out, the swim jigs produced for us again. I brought the tactic back to Southern Illinois with me, and Dad and I immediately started catching muskies with it. Especially in vegetated areas too thick for most presentations. The technique works great for muskies, until the season winds down for summer.

Speaking of summer. The late spring transition period and early summer sees the fish changing behavior yet again. Some muskies remain in shallow cover until it gets really hot. And we catch them on the same swim-jig-and-shad combos, as well as the multiblade spinnerbaits. This is also one of the two short time periods of the year when inline spinners can produce well. The safety-pin-style spinnerbaits are top options all year round here. They almost always outproduce the classic inline spinner designs so popular in the northern states, often called bucktails. But for a short period of time, as waters are warming fast, inlines can produce good numbers of SI muskies as well.

This is also a time of year when a unique and almost bizarre pattern can materialize for muskies here. Chris Shannon and I have caught some huge marlin and sailfish in Central America while high-speed power trolling with topwater lures and big live baits. The most common way to catch billfish, it's also effective for dorado, wahoo, tuna, and other giant offshore pelagic predators of the sea. Still, I've never heard of anyone using these methods in fresh water. It's not usually the best way to catch muskies, but we have had success when the bite is slow. It produces especially well when skis are spread out across large mid-depth weed flats or weedy break lines. And this is key. It happens where some of the weeds come very close to the surface but very few mats are present in the area.

The trick is to learn the spots in the winter when the weeds have died off, to make sure there aren't any stumps or rocks to hit with the boat. We often trim the outboard up, just enough to keep the prop under water, and place a couple of big surface lures behind the boat. We also hold the rods in our hands instead of placing them in holders. We put one lure right behind the boat, 10–20 feet back, and the second at least double or triple that length. Big propbaits and buzzbaits both perform well in most situations. But buzzbaits, with a pair of single upturned hooks, get the nod over treble-hooked lures when weeds are taller or thicker, or when there's a lot of cut-off vegetation floating on top.

A large volume of muskies makes a major shift at this time too, and as the season progresses. More and more of them head offshore, to suspend in moderate depths over deep water. Like early winter, this is the second of two times of the year when trolling is usually the best option for muskies. As these postspawn fish attempt to regain weight they lost during the spawn, they gorge on shad and shiners that head offshore. We catch excellent numbers of big muskies trolling small crankbaits, from about 3 to 5 inches and 3/4 to 2 ounces. We troll them near and through schools of suspended gizzard and threadfin shad and golden shiners. This is also a time when vertical jigging can produce some suspended muskies with spoons, vibrating blade baits, and single-blade spinnerbaits. But trolling usually remains the top option. These patterns continue until we stop muskie fishing for the summer.

This is a good place to mention leaders. Leader selection is pretty simple most of the time. Just do the opposite of what most muskie anglers did until the 2000s. That sounds like a joke, but it's really not. An old standby, the multistrand braided steel leader should go the way of the dodo. Sure, they still work, but so does a rotary dial landline telephone. Actually, I'm not sure about that one. There are some applications for single-strand wire leaders, however, especially shorter lengths. These may provide better action with certain lures, like specific models of jerkbaits and some small muskie cranks. Matching some of these lures with light single-strand leaders, as small as 4 inches and 20- to 30-pound test, can be needed to get the proper action out of the bait. But remember that softer rods and more care are required to land these fish without a break-off. I began buying offshore-grade fluorocarbon leader material from saltwater fishing catalogs many years before it became mainstream. With snaps and swivels, we tied our own leaders, and still do today.

Nowadays, most anglers in the know fish fluorocarbon leaders most of the time. And many commercially made models are available. Fluorocarbon is extremely tough material with high abrasion resistance. In thick, high-pound test ratings, these stack up well to steel. They might not offer quite the cut-off deterrence of steel, but we're fishing for muskies here, not barracuda. And what's more is that in all the years Dad and I have been chasing muskies recreationally, and then professionally, we've only had one single bite-off from a muskie with fluorocarbon. At the same time however, we've lost over a half dozen muskies from broken multistrand steel leaders. Fluorocarbon offers additional advantages. Fluorocarbon is a clear material, with virtually the same light refraction index as water. This means that light rays pass through fluorocarbon like they do through water, making the material invisible to fish.

With the exception of the single-strand steel wire that we use for just a handful of baits, there are just two other exceptions. Fluorocarbon is dense and naturally sinks. So, for a small handful of specific slower-moving topwater lures, like walking cigar plugs and big chuggers, we'll use a leader we tie ourselves made from saltwater-grade monofilament leader material. This is a strong and thick material, with higher abrasion resistance than regular monofilament fishing lines. It naturally floats well on the surface with topwaters. We use 60- to 130-pound test to tie leaders from about 10 to 15 inches, often attaching directly to the main line with a double uni-knot to eliminate the swivel. The other exception is for structure trolling. We don't do a lot of it. But in cases when we're trolling deep-diving crankbaits on the bottom, we'll often go to a titanium leader. Snapback titanium wire leaders are about as indestructible as it gets. Like fluorocarbon but unlike steel, titanium leaders do not kink. They're incredibly strong, and the most durable material available. But they are not invisible. The hope is that in deep water, with a moving bait, muskies just attack rather than follow to get a good look.

When digging into the structure, banging bottom and slamming into rocks and wood, a titanium leader will take the abuse, spittin' and cussin' along the way! In this instance, we make a long leader to protect the line. It's usually about 30 to 36 inches of 75- to 100-pound test wire. Day in and day out though, for probably 95 percent of our muskie fishing, we're using 80- to 100-pound test fluorocarbon leaders, with snaps and swivels. We may jump up to 130-pound test for oversized lures and heavy cover. We also drop down to 60-pound test fluoro for the downsized muskie and pike lures. For casting, with some presentations we might go as short as a 6-inch leader or as long as a 20-inch leader. But about 12 to 15 inches gets the nod most of the time.

For the second half of our muskie year, we get back after these big tough predators just as soon as waters cool down significantly in late summer. Once waters get back to around 80 degrees, we can successfully release these fish again and the best muskie angling of the year gets rolling. The hottest tactics for muskies remain pretty similar from late summer all the way into late fall. Fast fishing and aggressive retrieves are the norm most days, but small to medium-sized muskie lures are usually best. Once again, long-arm multiblade spinnerbaits remain our number-one option, in sizes ranging from about 1 to 3 ounces.

Shallow-diving floating crankbaits and minnow baits are a top choice now too, in sizes ranging from about 4 to 8 inches in length. Andrew Adank also loves fishing crankbaits for muskies. I netted his first toothy tough guy, a nice fall Southern Illinois fish of around 15 pounds that came on a 7-inch shallow diving jointed plug. The skirted weedless swim jig is also a great selection now too, in a head weight ranging from 1/2 to 2 ounces and paired with swimbait trailers from 4-1/2 to 7 inches. Paddletail versions usually work best. But we've had some success with segmented forked-tail designs, curly-tail grubs, and even reapers too.

Surface baits are better for muskies in fall than at other times of year, and we get a good consistent topwater bite for several months. Buzzbaits, propbaits, frogs, and walking-cigar-style plugs produce the hottest topwater action in the cooling waters, from 4-1/2 to 8 inches and 3/4 to 3 ounces or so. There seems to be a short time period in late summer when inline spinners produce as well as safety-pin spinnerbaits in similar weights and sizes. Additionally, while they can produce an occasional fish at any time of year, big wide wobbling spoons from about 1 to 3 ounces seem to work better in mid- to late fall than any other time of year. Wide-bottom versions swing wildly from side to side, while creating tremendous flash and a unique action that muskies seem to find attractive in cooling waters.

We don't discuss live-bait fishing much for muskies, because artificial lures work so much better most of the time. But we have caught muskies on various live baits over the years. Some anglers who travel from up north do it in Southern Illinois too, primarily in late fall. The reason why I mention northern anglers is because live bait fishing, sucker fishing to be more precise, is popular for muskies across the northern states. And many bait-and-tackle shops in the north carry giant suckers. There are no bait shops in Southern Illinois, that I've been able to find anyway, that carry large suckers for bait.

For muskies, suckers usually range in size from 6 to 14 inches. But anglers often use the biggest ones they can find, sometimes dragging massive muskie suckers that are more than a foot and a half long! These are expensive to carry and to buy, and it's mostly a fall season thing in the north. Terry and Janet Graeff brought down an order of suckers from up north one fall season to Top of the Hill Bait Shop. Unfortunately, they just didn't sell well enough to justify doing it again. But traveling anglers will sometimes bring the big baits down in the live wells of their boats to fish with, although mostly with limited success.

While we do have suckers in some SI muskie waters, they're probably not the same species as most northern baits. And suckers are not the dominant forage in our lakes anyway. We've tried the largest shiners from local bait shops. We've caught an occasional muskie on them, but they're not big enough to truly select for the species. We've also used the big goldfish that some bait shops sell, primarily to river catfish anglers. But we've had our best success using live baits that we catch or trap in local waters, such as chubs and bluegills. We've caught a few smaller muskies by accident on medium shiner minnows while fishing for crappie, and on large fatheads fishing for bass. But the chubs and bluegills have worked best. These have produced muskies over 40 inches, usually with slip-sinker rigs on the bottom or under floats fished up high, on jigheads with 4/0 to 6/0 hooks or on 5/0 to 8/0 circle hooks.

One last thing to mention with regard to top tactics for muskies is how to handle following fish. Because muskies routinely follow lures to the boat, they get a good look at them. They can do this because they are the big dog on the block, the apex predator. Like a polar bear or crocodile, they have little to fear by leaving the security of a group or the security of cover. A large enough percentage of muskies follow lures to the boat that muskie anglers have adopted the figure-eight technique. We watch our lures, or our line when lures are running too deep to see. As a muskie approaches but before our lure gets to the boat, we sink our rod tip in the water. Without slowing down, we turn to the side and then draw a large figure eight in the water. Sometimes a big oval will work.

This simulates a prey item making rapid directional changes to try to avoid being eaten by a predator. And it can drive muskies wild. Like the old cat-and-mouse scenario, the lure ambles along while the muskie follows. And when it tries harder to get away, the muskie strikes. In Southern Illinois, we probably catch at least 10 percent of our muskies on the figure eight. But sometimes it's a quarter of our fish or more, depending on lures, water clarity, and other factors. When a muskie follows but does not strike, we'll continue with the figure-eight maneuver as long as the fish follows or remains close. While Dad and I were on an *Adventure Sports Outdoors* TV shoot with Harry Canterbury, I had a muskie follow my School N Shad spinnerbait to the boat. The water was clear enough that you could plainly see the muskie chasing the lure on camera. The fish tried to eat it but missed. It swiped at the bait over and over again, and actually followed for over five minutes before we lost sight of it.

We performed a quick change-up. This is a common tactic of quickly switching to a different lure, usually something similar, but smaller or more subtle. You just make a few casts to the area where the muskie came from and the direction it swam off to. And sometimes it will strike the second offering. You don't want to sit on the spot and pound it for a long time. Just make a few casts and move on. We didn't hook that one. But we caught other muskies that day on the spinnerbaits, along with some big bass, and then returned to the fish later. You should always consider returning to a good fish that followed earlier in the day. When we came back to that one, we weren't able to catch it, but it does happen often. With muskies, you're working with so few fish per acre of water that when you locate one, you give it your best shot.

Smart muskie anglers will return to good-sized fish later in the day that had previously followed or swiped at a bait. When returning later in the day, it's best to make a few casts with the lure it was interested in before switching to something different. It pays off often enough. The timing to go back on a fish depends. It's best to go back at least a couple hours later, during a light or weather change. The fish is likely to be in a different mood, and maybe more inclined to eat. Often, unless there is a major weather change, we'll wait until the end of the day to go back after the biggest fish we raised. And these are all common tactics of the muskie hunter.

What isn't so common is a tactic that Dad and I developed quite a few years ago while traveling on the tournament circuits. We coined it *the turn and burn*. I later wrote an article about it for *Esox Angler* magazine, and other anglers started using it to put additional muskies in the boat. After many years of guiding and competing, with so much time on the water, we've been able to try out a lot of unproven ideas that we've wondered about over the years. While most anglers will leave a good-sized muskie alone after the change-up and return much later, we do the same. But before that, we often give the fish one more shot, especially if it's a big one. And it has paid off for us big time. This depends a bit on the attitude and behavior of the following fish. But more often than not, it's a good bet. After the following fish swims off, we make the change-up. We give it just a few casts and then move on. We wait a short period of time. We'll fish on down the structure and usually come back to the fish in just 10 to 30 minutes or so, even with no change in light or weather. That's the *turn* part of the tactic.

When we come back through, we don't use the lure the muskie followed, and we don't use anything similar to it. We pick a different style of lure. Any style of lure, as long as it fishes extremely fast that is. And that's the key, speed. We come back through the area and blast the spot with casts from a high-speed lure, and we fish it as fast as we can. That's the *burn* part of the tactic. We burn lures through the spot at warp speed, in an attempt to trigger a reaction strike. Basically, we know where a muskie is. We know that it's active enough to follow, but not quite enough to commit to eating. So we play on its instincts and force-feed that freshwater cuda!

Lipless crankbaits from 3/4 to 1-1/2 ounce, small heavy spinnerbaits from 1 to 2 ounces with willow leaf or hatchet blades, or small but heavy swim jigs from 1 to 2 ounces with downsized trailers are top choices. Another option is a shallow-diving minnow bait from about 5 to 7 inches, usually in a saltwater model that will run shallow but stay true at high speeds. We don't use any trailer on the spinnerbaits to reduce water resistance and allow it to be burned faster, and we'll also trim and thin out the skirt material. We trim and thin skirts on swim jigs too, and use skinny trailers with minimal bulk that produce little resistance to increase speed. Longer lighter-weight rods with moderate fast actions are best when paired with fast reels with high-speed gear ratios of 7:1 or faster, allowing anglers to set the water on fire. And that's what makes the turn and burn so effective. Many times, we don't have to go back on a fish at the end of the day because we force fed it earlier.

Gone are the days of fishing with pool cues. When I first started muskie fishing, I actually had one jerkbait rod that was just 5-1/2 feet in length. It probably had something along the lines of ultra-heavy power and an extra-fast action. Longer rods facilitate longer and more accurate casts. They set big hooks into boney jaws at extreme distances. They handle big powerful fish well during the fight. And they allow anglers to actually land the majority of fish hooked, with some skill. While we do occasionally use something a little shorter or a little longer, 95 percent of the muskie fishing we do these days is with rods from 7 feet 9 inches to 9 feet in length for casting, and 8 to 10 feet for trolling. Trolling rods are softer, medium to heavy power with moderate actions. Casting rods are stiff and range from medium-heavy to ultra-heavy power with moderate fast to extra-fast actions, depending on the application.

We use big round bait-casting reels in line-counter models on all of the trolling rods. But we select the low-profile wide-spool bait casters for all casting rods. The newer designs allow for plenty of capacity with heavy lines while being compact and sleek enough to palm the reel, for all-day casting comfort. We prefer standard handles for most reels but replace them with power handles on the ultra-heavy power rods, used exclusively for oversized lures. As for lines, on most reels we spool a super braid in 65-pound test for small lures and 80-pound test for standard-sized baits. We drop down to no lower than 50-pound test for downsized muskie and pike lures and step up to 100-pound test for large and oversized muskie lures.

Catfish

Tactics for catfish are much simpler than those for many other species, because artificial lures are not really a part of the equation. Yes, we do catch cats on lures here. But natural baits are just so much more effective that we seldom target catfish with lures. Crankbaits probably take more catfish than any artificial because channel cats will eat them with more regularity, casting or trolling. Channels make up the vast majority of catfish catches on lures. Blue cats are least likely to take an artificial bait. But flatheads and bullheads both take lures, including crankbaits. We've caught some nice cats on spinnerbaits and vibrating blade baits. We've also caught quite a few vertical jigging with spoons, especially in late summer.

I actually landed a 15-pound Southern Illinois channel cat on a topwater lure one time. While fishing at night for bass in the heat of summer, I did not see but heard a loud strike shortly after my lure gurgled its way away from a rocky shoreline bank. The fight in the dark was epic. Jigs with scented soft plastic lures are probably the second-best producer of cats with straight artificial presentations, and this tactic can be easily modified specifically for catfish. We have had some success using painted plain jigheads and weedless skirted jigs for catfish. We add more scent to a scented soft plastic trailer at times, but often replace it with a live bait instead.

Most often, we lip-hook a slender baitfish species like a shiner or fathead minnow, a sucker or a chub. Lively offerings usually work best. But you can also use dead baits if that's all you have. And you can use the rod to manipulate the jig and bring the dead back to life. Jigs allow an angler to add some color and bulk to a catfish offering. Goldfish, which are much fatter, as well as bluegills and other sunfish species, which are tall and deep-bodied, can work too. When fishing a live bait on a jig by itself, you can cast the offering out and still fish in one spot, especially if there is wind or current. Casting and retrieving the offering works well for covering water, although a slower retrieve with pauses is usually best.

When repeatedly casting and retrieving the offering, some live baits are poor choices. Larger baits usually work better than smaller ones. Shiners and fathead minnows have fragile mouths and will often tear off with repeated casting. Suckers, chubs, sunfish and similar baits however have hard, tough mouths that can take the abuse. Still, it never hurts to add a couple of spinnerbait trailer hook keepers, or pieces of shrink tubing or surgical tubing on the hook. This largely unknown trick helps keep the bait from coming off during a cast.

Another artificial and live-bait combo that works well is a spinner rig, also called a harness. These rigs use a spinnerbait blade, spinning freely on the main line with a clevis. They feature a series of beads to add flash and color. A single-, double-, or triple-hook system, using everything from night crawlers, red worms, and leeches to suckers, shiners, and other minnows, completes the business end. These are used in conjunction with a slip sinker and slowly trolled and dragged on and near bottom. It's a deadly tactic.

Down to the meat and potatoes of catfish tactics. The methods that work best most of the time are classic catfishing rigs and baits that have stood the test of time. The traditional slip-sinker cat rig, also referred to as a Carolina rig, is what we use most often. With a sinker sliding freely on the main line, we use a quality barrel or crane swivel, along with a leader and the hook. Be sure to add a bead on the main line ahead of the swivel to protect the knot from being damaged by the sliding sinker. With a molded-in swivel, the bell sinker is what we prefer most of the time. An egg weight can also work. When current is fast, a thin heavy teardrop-shaped weight will lie flat on the river floor and stay in one place rather than rolling and washing along a swift bottom area. Leader length varies. Most of the time we use a leader about 12 to 18 inches in length. A longer leader allows live baits freedom to move around and get noticed a little better. But we often shorten up to just 4 to 8 inches when fishing heavy cover that our baits can get snagged on easily.

The Kentucky rig has become our number two choice. Here we use dropper knots to tie small leaders from about 4 to 12 inches directly with the main line. We then attach a heavy bell or bank sinker to the bottom. By tying loops into the tag ends, anglers can change weights and hooks without retying. Much like a double drop-shot presentation, this allows an angler to present two different offerings to cats at two depths, except with more wiggle room for the baits. I'll usually place one dropper a foot or so above bottom and the second 2 to 4 feet up above that. Sometimes when fishing in fast current, it's beneficial to tie a swivel above the rig. But we tie direct for lake fishing.

For suspended cats, we attach a large cigar- or pole-style slip float sliding free on the main line with a bead and bobber stop above. We use an egg sinker sliding on the main line with a bead and swivel below, and a leader about 10 to 18 inches to the hook. These three rigs are our go-to choices for flatheads, blue cats, and big channels. To keep things simple when targeting smaller channel cats and bullheads, a simple hook and split-shot weight can be fine for bottom fishing. Or a small cigar- or pencil-type slip bobber, or a fixed float with split shot will work to suspend smaller baits.

For bullheads and smaller channel cats, we've used the larger Aberdeen hooks when we want a long shank. But short-shank live bait hooks and mid-length shank baitholder hooks are our go-to when choosing J styles, in sizes ranging from #4 to 2/0. Treble hooks are best with stink baits and gobs of worms. But these are typically only used in smaller sizes from about #8 to #1. For bigger channels, we do use some standard J hooks in strong heavy wire models like the O'Shaughnessy when we need a long shank. But again, short-shank live bait hooks and mid-length shank baitholders are the go-to when we're using a standard hook style. Just in stronger and heavier wire versions, in sizes ranging from #1 to 5/0 depending on bait size. However, we've transitioned mostly to circle hooks for big channel cats the last few years. Typical sizes range from 3/0 all the way to 8/0, depending again mostly on bait size.

Circle hooks are almost always the best choice for catch and release, and the safety of the fish. They slide out of the throat and almost always hook fish firmly in the corner of the mouth. But they offer other hooking advantages too. With standard J hooks, knowing when to set the hook is important. With circles though, you leave the rod in the holder and allow the fish to hook itself. Then just pull tight and start reeling. Although some anglers swear by offset circles for better hookups, we've had about equal success with offset and standard circle hooks. Some tournaments no longer allow offsets. And if you're practicing catch and release, standard circle hooks can be removed from the fish easier and with less damage.

But our big game changer with circle hooks has been in the gap. We use a variety of standard circle hooks, double-action circle hooks, and octopus-style circle hooks. But they're primarily in wide-gap and extra-wide-gap designs. When targeting big fish with large live baits or large pieces of cut bait (our bread-and-butter tactics), the wider-gap circles really shine. They provide more room between the bait and the hook point and allow for a better bite on big catfish for increased hook-ups. And this takes us to the giants. For flatheads and blue cats, we use even bigger and stronger hooks. For some fishing situations, we still use standard J-style hooks like the long-shank O'Shaughnessy. As well as heavy-wire short-shank live-bait hooks and medium-length octopus-style wide-gap J hooks.

Over the years, we've switched primarily to circle hooks for flats and blues too. We use the same various styles of circles that we use for big channels. And again, all in wide-gap or extra-wide-gap designs. Heavy wire hooks for these beasts are the norm, in sizes ranging from 6/0 to the giant 12/0 circle hook. In waters that have the capability to produce giant cats from 50 to 100 pounds or more, bigger, stronger hooks are best. But this is about bait size more than anything. The bigger the bait used, the bigger the hook required. With huge offerings, the hook has a lot of meat to go through. You need space, with plenty of the hook shank exposed past the barb for good hookups.

For river fishing, we usually hook live baits up through the lips. This way they can swim with the current and stay lively and natural looking. In lakes, we primarily hook them in the back, between the dorsal fin and the tail. Usually about one-quarter to a third of the way up the fish's body, from the end of the tailfin. This is where the large wide-gap circle comes in. It exposes plenty of steel with which to grab the inside of a big cat's mouth nice and solid. The same principal is applied with big cut baits. Do not hook them through the middle but rather in a corner, or in the side of the bait. Just make sure that most of the hook is exposed. Dad and I have spent many years catfishing the mighty Mississippi and its tributaries. When we started, information on trophy catfish tactics wasn't readily available like it is now. And we learned mostly by trial and error. We've primarily used the same J-style hooks for years, and mostly with night crawlers, grubs, grass-hoppers, and crayfish.

The first big catfish I ever caught on another fish was a double-digit channel cat from the private lake on our family farm. I took that one with a J hook and a solid hookset. But I missed several others before-hand, likely because there wasn't enough of the hook exposed. Still, Dad and I began using bluegills on some bigger lakes. And also on the Mississippi River and its feeder streams, but with limited success. Burying the hook point in worms, insects, stink baits and other soft offerings worked great, but not with baitfish. Eventually, we began exposing the hook point just a bit, and with it increased hookups. We experimented further by leaving the entire barb exposed, and we landed more fish that struck. Eventually, we figured out that hooking the baits just enough to keep them pinned on during a long cast was all that was necessary. And our catches went higher than ever with many types of J hooks.

One of our favorite fishing buddies from way back in the day, my cousin Brian Duffey, went to college in Florida and had done some saltwater fishing there. Circle hooks had been popular with saltwa-ter anglers for many years before they saw regular use in freshwater circles. Catfish feed much like sharks. And after talking with Brian about saltwater rigs, we started employing circle hooks for catfish. In the beginning, we had little success, and I kind of lost interest in them. After I'd been fishing in salt water for a number of years though, I'd used circle hooks with good success on a variety of species. But it wasn't until I'd had some successful trips for big sharks that I saw just how good circle hooks could be for big catfish, if used properly.

The three main keys are hook and gap size; hooking baits with the hook point, the barb, and plenty of the shank exposed; and allowing the fish to double the rod before pulling back, to ensure that it's hooked itself solidly. It does feel funny at first to grab a rod a fish is on and not set the hook, but losing a couple of fish will break you of that habit pretty quick. The more we experimented with circles, the more fish we caught. But as time went on, we began to select for larger catfish with larger baits. And the big wide-gap circles became more and more important to our success.

Top tactics can vary a bit depending on catfish species. There are no hard and fast rules, and any of these baits and methods can take any of these fish. But we break things down based on the type of fishing that we're focusing on for a particular day. For channel cats, just about anything you can think of will catch them at times. Anglers looking for fast action and a chance to load the boat with cats for a fish fry often choose baits and methods for small to mid-sized fish. Commercially produced stink baits are very popular with channel cat anglers. Fiber doughballs or nuggets work best when molded onto bait-holder catfish treble hooks. These feature a curly wire running the length of the shank to help keep the bait on.

A tip that few catfish anglers have ever heard of is using a feather treble instead. Feather trebles are primarily used on the back of poppers, jerkbaits, and other lures designed for bass or walleyes. While these don't hold the bait on quite as well, the hair, thread, glue, and other materials on the shank will hold bait better than a plain treble alone. The feathers, bucktail, flashabou, sparkle hackle, or other materials designed to attract and trigger more bass strikes add a bit of bulk, color, and flash to the bait. And, protruding out from the bottom of the stink ball, these materials will wiggle with the slightest wind, current, or other movement underwater. When cats are actively feeding it seems to make little difference. But on tough days, we've seen these outproduce plain trebles significantly.

Other commercial stink baits that work great for Southern Illinois channel catfish include liquid baits, squirted into plastic teardrop cat-fish lures. And dip baits or punch baits work best with sponge trebles, catfish tubes, or mesh-bait holders with treble hooks. Some anglers mix up their own stink baits at home, which can work well if you don't have neighbors close by. Night crawlers, red wigglers, grubs, and other worms work well for channel cats. Crickets, grasshoppers, horseflies, and pretty much any bug can catch these fish. Shrimp and hot dogs are popular. And crayfish, frogs, salamanders, mice—you name it, and channel cats will eat it. The hot bait that's gained a crazy amount of nationwide popularity, chicken soaked in strawberry drink mix with garlic salt, is a staple for many channel cat chasers now. I know, it sounds strange. But many anglers swear by it for numbers of channels, and they're also soaking beef, fish, and all kinds of other baits in the stuff too.

All baits can produce channel catfish all year round. However, those previously mentioned are often not the best for big channel catfish. They can and do take the occasional trophy channel. And they are often the best bet for fast action, from high numbers of these whiskerfish. But there is no question that most larger channel cats prefer to eat other fish most of the time. Big channels readily scarf up both live and dead fish. Whole dead baits can work at times. But cut bait is a much better option. It disperses more scent into the water, attracting feeding catfish from greater distances.

Live fish work great too. Small, medium-, and large-sized shiners and fathead minnows from local bait shops are a top choice. Some dealers carry goldfish in small and medium sizes that can work well too. We've run our own homemade baitfish traps in streams and lakes for many years, but commercial models will work. All of the chubs, suckers, darters, silversides, and other various minnows found in local waters are excellent baits, from about 3 to 8 inches. We've used cast nets to catch gizzard and threadfin shad, skipjack herring, golden shiners, and other pelagic baitfish from lakes and rivers. These produce lots of big channel cats in similar sizes too. The most fun way to catch our bait is to go fishing for them. Bluegills and other various sunfishes are a top bait choice for big channel catfish throughout much of the year, from 2 to 7 inches. The best way to catch these feisty little game fish is on rod and reel. As a plus, big golden shiners, gizzard shad, and skipjack herring can be taken this way too.

Bullheads are the smallest catfish species we target in Southern Illinois. We definitely catch the most bullheads on small and medium-sized live shiners and fathead minnows than anything else. Dead baitfish can work, but lively ones outproduce. The smaller sizes of chubs, suckers, and darters work great too. And we have taken a few over the years on the smallest bluegills and sunnies. The number-two choice behind baitfish would be small to medium-sized crayfish, which bullheads love. Grasshoppers, crickets, and other insects will take these cats too. Night crawlers, grubs, and red wiggler worms produce fish. And we also catch them on the smallest worms, like mealworms and wax worms, typically reserved for bluegills and other panfish. Bullheads will sometimes bite stink baits, hot dogs, and the like, but they really prefer live offerings.

On to the true giants. It actually gets simpler with the biggest catfish. These opportunistic predator-scavenger fish can be caught on most items previously mentioned at one time or another. But flatheads and blue cats are far more selective than channel and bullhead catfish during most of the year. Flatheads behave much like the largest channel cats do when it comes to feeding. It's the seafood buffet they prefer. Younger flatheads love crayfish and eat them heavily, along with baitfish and some small game fish. Adult flatheads still eat crayfish too. But baitfish, game fish, and rough fish make up the vast majority of their diet on most waters, all year round. Cut bait does produce flatheads, especially from winter through early spring, as well as following the spawn from late spring to early summer, depending on water temperature.

But day in and day out, big baitfish, game fish, and rough fish are the number-one bait choice for the king of the water. Live bluegills, redears, pumpkinseeds, greenies, warmouths, and all of the various sunfishes are a top choice for flathead catfish most of the year, in sizes ranging from about 4 to 10 inches. Big suckers, chubs, darters, shiners, shad, herring, and similar baitfish species also take brutes in sizes from about 5 to 20 inches. Although some anglers have reported landing big flats on giant baits over 2 feet long. Speaking of giant baits, flatheads will devour rough fish too. Common carp, grass carp, bighead and silver carp, drum, buffalo, and even predators like gar can all fall victim to a big flathead, and can be good bait options if you catch them. We'll rig up carp to about 3 pounds if we catch them. But anglers have reported landing huge flatheads on various rough fish weighing 5 to 6 pounds or more!

It's quite a bit simpler when it comes to blues. Blue catfish eat cut bait so well, that there's rarely a need to try anything else. In fact, if a group was fishing a lake or river with no catfish in it except blues, and a dozen rods were put out, 11 of them better have cut bait. Any kind of cut bait can work for adult blues, from bluegills to warmouth and from carp to gar. But fatty, oily baitfish are like steak and lobster, as good as it gets. Gizzard shad, threadfin shad, golden shiner, mooneye, and skipjack herring are top options. If these aren't available, suckers or chubs will work. In most cases larger pieces of cut bait are best. Besides leaving a better scent trail with more oil, guts and scales in the water, cats can sense a larger chunk of bait lying on bottom or moving in the current. Blue cats will eat huge cut baits. In the instance when you can't catch large baitfish, use multiple small ones on a large wide-gap hook. Pack on as many baitfish as you can, as long as the hook is exposed well. Just be sure to cut a couple of them, to get more scent out in the water.

A few tips that make catfishing more productive include the use of lighting. Submersible lights can be dropped down below the boat or below a dock at night, causing the water to glow green. This attracts flying insects to the water's surface. The underwater lights draw in plankton and aquatic insects. These bring in shad and other baitfish, which of course then attract all predators, including catfish looking for a meal. Lighting above the water can work too. Long before the time of Edison, tribal peoples from the Achelous to the Amazon, the Mekong to the Mississippi, the Zambesi to the Zeya have used light to draw fish for nightly catches. They built fires along the water's edge to activate the food chain, a technique that can still work to this day. Lanterns work well too. And fishing below lighted docks can be highly productive as well, especially when it's the only significant light close to the water in the area.

European anglers have long used a fish call, called a clonk. They plunge the device into the water, which creates a bubbling sound similar to uncorking a bottle. This increases the chance of catching the monstrous wels catfish in their rivers. Saltwater striper anglers often bang a mallet on their boat decks off the northeast coast to attract fish. We routinely make splashing sounds on the surface of the water, just before we start fishing for sunfish. We do this to draw more sunnies into the area to investigate the sound, and it works for catfish too. I turn my hand upside down and slap the surface with the back of my hand. I create a popping sound, by striking the water hard and immediately pulling my hand back. I make several pops very quickly which sounds a bit like fish feeding competitively on the surface. It's easy to see the sunnies get cranked up since they'll readily come to the surface.

But we've had catfish actually come right to the top as well. And it does seem as if we catch more catfish, even off the bottom in deep water, when we regularly pop the surface. Another tactic that I like to employ is to keep a rod with a rattling crankbait on the deck when catfishing. It often seems as if the catfish bite a little better when a noisy lure is pulled through the water now and then. We'll usually make a few casts and then put the rod down, wait a while, and make a few more casts, so it's not constant noise. But the rattles mimic clicking sounds that baitfish and crayfish make in the water, and can help draw the cats in.

Chumming is the best method of all for attracting catfish, and it works great for all species all year round. Some anglers make their own homemade chums. But there are various commercial options that make things a little easier. Handfuls of loose grainy chum can be thrown out into the water in different directions around a fishing spot. This material floats on the surface or sinks very slowly. Then we take some of the stink bait doughballs, which can be thrown a greater distance, and hurl them around the area where they quickly sink to the bottom. Lastly, we use a liquid spray-on scent, one with either a shad oil or crayfish oil base. We spray that on the water around the boat, which creates an oily slick. This way, the fishing area is covered from top to bottom, with strong scent and taste particles that not only draw fish in from long distances but also get them in a mood to eat.

Tackle is a lot simpler when it comes to catfish too. With muskie fishing, heavy powerful rods are required, but they must be lighter in weight to be comfortable to cast with all day. Bass rods should be extremely sensitive to detect subtle bites when fish aren't active to allow you to react in time to hook the fish. Trout rods have to have the right action to cast very lightweight lures adequate distances. For catfish however, it really just comes down to strength and durability. Sure, all rods for all species should be strong and durable. But you sacrifice some of that when factoring in other important qualities. When choosing catfish rods, select strong durable sticks, and you're pretty well good to go. The actions of catfish rods are mostly going to fall into the moderate or moderate fast range, since you want the rod to bend more evenly over most of its length. This lends well to handling powerful fish.

Our smaller catfish rods for bullheads and smaller channels are light to medium power from 6 to 7 feet. The rods we use for flatheads, big channels, and blue cats are medium-heavy to heavy power, from 7 to 9 feet in length. The longer rods go on the outside rod holders to help spread baits apart well. We use a few spinning reels with large spools that hold plenty of heavy line. These can be a little better for making extra-long casts. Most of our catfish reels, however, are round wide spool bait casters with clickers. For bullheads and smaller channel catfish, we usually spool 10- to 14-pound test monofilament on spinning outfits, and 14- to 20-pound test on bait casters. We choose lighter line in clear-water lakes without a lot of abrasive cover but opt for the heavier stuff in swift rivers and areas of lakes with a lot of snags.

For big channels, blues, and flathead catfish, we usually spool 30- to 40-pound test mono on big spinning reels and jump up to 50- to 60-pound test on bait-casting equipment. We have used braided lines from 50- to 100-pound test in some instances. And you'd assume that since it's our go-to for most muskie fishing, it would be for big cats too. But muskie fishing is faster and usually higher in the water column. Targeting cats, however, is mostly still fishing, or moving slowly on and near bottom around thick snags. Heavy monofilament has been the best choice for us, because it's so thick in diameter and because it also has a bit of stretch for shock absorption.

Be sure not to skimp on terminal tackle either. Use swivels and leaders that are at least the breaking strength of the main line or higher. We like to use stronger leader material and swivels than the main line for big cats. This is because the long main line can stretch and take more shock, in a way that a short leader or swivel can't. Some rods have glow-in-the-dark tips, which are great for night fishing. But battery-operated lights and glow sticks can be attached to the end of a pole to aid with vision too. Clip-on brass bells are attached to the ends of rods that alert anglers to bites. And electronic sensors that beep when line goes out are also available to attach above reels.

There's one last tactic I'll mention here that can take big catfish at times. And sometimes when they're just not biting, this technique can save the day. It's called mooching. A number of years ago now, I took Mom and Dad up to our outfitting company's new charter fishing resort location, right on the shores of the northern Pacific Ocean, in Craig, Alaska. After arriving in the southeast portion of the state, we took a small plane out of Ketchikan to Klawock, on Prince of Wales Island. From there it was a short trip by car, to where we spent one of the best weeks of fishing in our lives. We settled into our second-floor rooms, with an incredible view of the coast just a rock's throw away. While over a dozen giant bald eagles perched in tall evergreens, literally within arm's reach of our windows. They wait every day for the return of the charter captains, and the scraps from cleaning the day's catch.

After a fresh seafood dinner on the deck above the water and a good night's sleep, we rose early. The three of us hopped in the twin-engine offshore boat with our new CSO pro staffer, Captain Rafael Ramirez Ruiz, and headed out of the bay. Although we fished mostly in tee shirts, the giant rugged peaks of the Coast Mountain Range surrounding us were still snowcapped. A place known for wildlife, we saw blacktail deer, huge black bears, giant sea otters, a variety of seals and whales, and a rare Alexander Archipelago wolf. Just offshore in the cold, cobalt blue North Pacific waters, a short distance from river mouths where salmon ascend to spawn, we grabbed heavy-power big-game rods Captain Raf had prerigged. The rig was something I had never seen before, which was quite unusual.

Raf explained the technique to us. We started fishing, and literally within minutes we were all hooked up! Dad was fighting a huge king salmon, while Mom and I were locked in battle with two big halibut. A colossal fight ensued, with Dad's fish rocketing up and leaping high into the sky over and over again. This while the halibut that Mom and I were on were trying to dig their way back to the bottom in over 200 feet of water! Mid-fight, a pod of humpback whales surfaced close to boat. I asked Cap to take my rod so I could grab my camera. He replied not to worry, we'd see more whales, which we did throughout the entire amazing trip. Eventually, all three of us landed the huge Alaskan fish we'd hooked up. It was more of the same for the rest of the week. And we brought back so many fish fillets we almost had to go out and buy another deep freezer when we got back home.

What was the technique that worked so well for us on these giant fish, and so many more throughout the week? Mooching. A funny name for a fishing method yes, but we were sold. While we used a couple other fishing methods that week, mooching worked so much better, that there was no reason to do anything else. And it didn't just produce our giant king salmon and monstrous halibut but also massive yellow eye, big black sea bass and other rock fish, and all of the huge ling cod we landed. During other times of year, it's a top option for the other various salmon species and sharks too. Mooching was incredible.

It's a fairly simple method of fishing that anyone can do, but the rig is unlike anything I've ever seen. A large, heavy banana-shaped sinker is tied into the main line with beads and a chain swivel. The chain swivel is critical to avoid the tremendous line twist that would occur without it. A fluorocarbon leader is attached below, which runs down to a double-hook setup. Two mid-length shank live-bait hooks are tied about 3 to 5 inches apart, depending on bait size. Then a whole dead baitfish, about 6 to 8 inches long with the head removed, is hooked. The head is cut off the baitfish at an angle. The first hook is placed in the side of the bait near the head with the hook point and barb exposed. The second hook is placed in the same manner about two-thirds of the way toward the tail.

Most anglers would look at this set-up and think, good grief, I can't fish like this, the bait is going to just spin. Exactly! It is designed to do just that, spin. The weight carries the bait down to the bottom, and the angler simply reels back up slowly. The chain swivel keeps the twist out of the main line. And the baitfish spins in a circle as it falls to the bottom, and again on its way back up. That is if it doesn't get hammered on the way down anyway, which it often does. We caught fish on both the drop and the rise, like it was magic, all week long. Adjusting the size of weight and bait, the same rig can work for freshwater species. But it is especially applicable for catfish here in Southern Illinois and has produced some big fish over the years. I wrote articles about this trip and the mooching technique for the *Murphysboro American* newspaper, and for *Adventure Sports Outdoors, Fishing Facts, Outdoor Guide, River Country Outdoors,* and *Simms Outdoors* magazines. But surprisingly, it's still a largely underused technique to add to a catfish angler's bag of tricks.

Sunfishes

Tactics for bluegills, redears, and all of the sunfishes of Southern Illinois are fairly simple. A wide variety of small artificial lures take all of the sunfishes. Small jigs made of hair, marabou, tinsel, feathers, or similar materials are effective in nearly all conditions. The same is true of plain jig heads matched with a variety of soft plastic lures. Tiny shads and other single-tail jerkbait and minnow-style lures; beetle tails and other double-tail plastics; hollow-body and solid-body tubes; small straight- and curly-tail worms; baby frog and crayfish imitators; chigger, hellgrammite, and other insect imitators; and single- and double-curly-tail grubs and spear grubs are all effective in sizes ranging from about 1 to 3-1/2 inches.

When casting and retrieving, jig weights can be effective from about 1/64 to 1/4 ounce, with 1/32- to 1/8-ounce jigs getting the nod for most applications. However, tiny jigs as small as 1/80 and 1/100 ounce are effective too. But these must be used in conjunction with a float, as these weights are too light for effective casting. Any size jig can be fished vertically however. Heavier jigs are needed for deeper waters. But even a 1/100-ounce jig can be vertically fished by itself below boats and docks or dropped into the shallows with a long pole. Panfish nibbles are small-scented bait balls that can be added to jigs and hooks if the bite is slow. And the same is true of squirt- or spray-on scents. All jig styles and plastics are effective for all sunfish year-round. It's important to note however, that anglers must match lure size to the species of sunfish they pursue, according to fish size and mouth size.

Bluegills, redear sunfish, longear sunfish, and pumpkinseeds for instance have very small mouths for their body size. So, unless fishing in waters known for numbers of larger-than-average-sized fish of these species, 1- to 2-inch plastics get the nod while 2-1/2- to 3-1/2-inch plastics are too big most of the time. Green sunfish, rock bass, and warmouth, on the other hand, have very large mouths for their body size. And while these species don't grow as large as bluegills and redears do, they often readily attack offerings much larger.

The other category of offerings, which produce all panfish throughout the entire year, are live baits. These can be presented on plain jigheads or as a trailer on the back of a hair jig or similar lure. Most commonly, we deploy them for sunfishes on short-shank live bait, mid-length shank-bait holder, or long-shank Aberdeen-style J hooks in sizes ranging from #12 to #4. A simple split-shot weight can be used to fish the bait along the bottom. However, a small drop-shot rig is a very effective presentation for fishing either one or two baits at a time. Especially in moderate to deep water. And the drop shot isn't utilized by a lot of panfish anglers.

Another great trick, and my nephew Cade's favorite, is to use small flies instead of plain hooks in your drop shot. A tiny bit of hair or tinsel adds color, flash and action to a live bait. And if the bait comes off, you've still got something that fish will bite. We've doubled up a lot on big sunfish of various species with live baits or nothing but flies. Another interesting tidbit is to add a squirt of oil-based scent. The fly materials hold the attractant well. It creates a tiny scent trail in the water that's even more pronounced than a live minnow or worm. Top live baits for sunfishes include mealworms, wax worms, red wigglers, as well as small night crawlers and grubs, crickets, grasshoppers and other insects, tadpoles, small crayfish, and small to medium-sized shiners and fathead minnows, silversides, or other diminutive baitfish. Natural baits are always a top choice for all sunfishes, 365 days of the year.

Gone are the days of the big round bobbers. Fish have to really take off hard to pull down one of these highly buoyant models. They're just not necessary in most situations, and many times are counterproductive. A smaller and thinner float is more sensitive and will aid an angler in responding to lighter-biting panfish, for increased catches. When targeting shallow cover, we mostly use small cigar- and pencil-style fixed floats to suspend our live baits under. For deeper water, we'll use a quality small slip float in the same thin shapes. Some floats are weighted. But most require a bit of added weight to stand the float up and make it easy for a fish to pull down. A piece of split shot or two is all that's needed to cap off the package and catch a cooler full of sunnies.

A few other artificial lures that tend to work best in the warmer half of the year include small spinnerbaits, bladed vibrating jigs, beetle spins, and inline spinners from 1/32 to 3/16 ounce and 1-1/2 to 3 inches. Tiny crankbaits and minnow baits from 1 to 3 inches and 1/10 to 1/8 ounce are productive, as are hard topwater lures like panfish poppers from 1 to 2 inches. Small casting and jigging spoons and tiny vibrating blade baits, from 1/32 to 1/8 ounce also produce sunnies for us. For most sunfish angling, we use ultralight power spinning rods from 5 to 6 feet in length, with moderate or moderate-fast actions for casting and jigging.

We do occasionally step up to light power rods from 5-1/2 to 7 feet, or down to super ultralight rods from 4 to 5 feet. Another exception are light long poles, or dipping rods, from 8 to 12 feet. We use open-face spinning reels probably 95 percent of the time. But we occasionally go to a closed-face spincast reel, particularly the underspin models for the long dippin' poles. Typically, we use monofilament lines from 2- to

4-pound test for sunfish. But we'll occasionally go up to 6-pound line for bull gills and big sunnies in heavy abrasive cover.

Walleye and Sauger

A variety of lure and live bait options produce marble eyes and gravel lizards all year round. Walleyes got their nickname because their eyes look like marbles, due in part to the uniqueness of the body part. The vision of the walleye is different than most fish, and they see better at night than most species do. Saugers and the hybrid saugeyes have similar vision. But the sauger got its nickname because it loves gravel so much and prefers to hold and hunt along gravel bottoms if available. It seems that most lures and baits will work, regardless of the time of year or temperature. But speed adjustments are more important for these species, season by season. When it's cold, we slow down, and when it's warm, we speed up, but most other factors remain the same.

Floating- and suspending-model crankbaits and minnow baits and sinking lipless cranks are top choices, ranging anywhere from about 1/8 to 1-1/2 ounces depending on baitfish size. Mostly, cranks from 2 to 4-1/2 inches or minnow plugs from 3 to 6 inches are selected, depending on available forage. In early to middle winter, both shallow and deep models can produce well depending on the depth of the fish. From late winter to mid-spring, shallow runners become more important as walleyes and saugers move shallow to feed and spawn. By late spring, though, many of these fish have moved out to deeper waters ahead of most other species. They don't necessarily hold in deep water yet. But they'll suspend over it, as well as on the mid-depth structures adjacent to it. And cranks are a great choice for this fishing.

Mid-depth running cranks and plugs can be fished down to the structures and out over the open waters next to them. Deep-diving crankbaits can produce well too, though on top of structures. They'll get down to the target zone fast and then bang into the bottom. They careen off the structure itself, and off of rock and wood cover objects lying on the lake or river floor at moderate depths. Banging lures into the bottom or into cover can be a great strike-triggering technique for all predators. Walleyes and saugers love it, especially with crankbaits.

When choosing cranks for this method however, it's best to choose highly buoyant floating models. These are less likely to get snagged. If they do get snagged, stopping the retrieve and giving a little slack line will often cause the floater to back out and up from the snag. If that doesn't work, trying snapping the rod tip on slack line hard and fast several times. This often breaks the lure free from what it's stuck on. And the buoyancy of the lure will bring it up and away from danger. When fishing the shallows, carefully reeling the rod tip down to the lure often frees it from the hang-up, with just a slight jiggle. But this doesn't work in deeper waters. Anglers should invest in a lure retriever to keep on board.

I have two styles. One is a long telescopic pole with a curly metal end that feeds onto the line and wraps around the lure. The other is my go-to, called a plug knocker. It's a heavy cylindrical weight with a curled metal wire welded to the side that wraps around the line. Several short lengths of light chain are attached to the bottom. This unit is affixed to a long section of strong cord. You feed the fishing line through the curly wire. And while keeping the rod tip up and the line tight, with pressure on the snagged lure, allow the heavy weight to slide down the line and crash into the lure. Most times, this will knock it free. If it doesn't, jiggling the weight causes the chains to grab the hooks. And you can often break the lure free, and usually only have to replace a hook. With the price of lures these days, this is a necessity. Many commercial lure retrievers are available.

We use custom-made plug knockers. They're about quadruple the size and weight of the commercial models, with longer, heavier chains and stronger cord. For manageability, we use the plastic float of a marker buoy to keep the cord wrapped and neat. In worst case scenarios when the lure will not come free, we'll wrap the cord around a boat cleat. We've used the trolling motor, or even the outboard, to break the hooks off and get the lure back. A few times, when hanging up a large muskie or striper lure with huge heavy-duty hooks, we've actually broken off submerged tree limbs or dragged logs up into the shallows with the outboard in order to get the bait back. Since many bass and walleye lures can top $15 each, with some muskie or striper lures ringing up at double to triple that amount or even more, it's well worth the effort. We save dozens of lures in various ways each year.

Crankbaits continue to produce walleyes and saugers through the entire summer period. In early summer we're still mostly fishing mid-depth structures and adjacent open-water areas. But in middle and late summer, many walleyes will use deep water as well. In fact, we often use the deepest running cranks to get down to fish holding around the thermocline. As late summer fades to early fall, these toothy fish start moving shallower again. They tend to stay offshore and use deeper water a little longer than most species. But once the majority of the shad and other baitfish have left these areas to invade the shallow confined waters, marble eyes and gravel lizards aren't far behind. As such, shallow and mid-depth running baits work well throughout fall.

Other top artificial lure choices for these fish include 1/16- to 1-ounce jigs, paired with various 2-1/2- to 6-inch soft plastics like single-curly-tail and spear-tail grubs, paddletail swimbaits, soft jerkbaits and other minnow or shad imitators, curly- or straight-tail worms, and marabou-tail soft plastic grubs, as well as tied marabou and hair jigs. Generally, smaller lures are better for sauger while bigger baits get the nod for walleyes and saugeyes. But it depends largely on size and available prey. Vibrating blade baits or jigging and casting spoons from 1/4 to 1 ounce work well too.

Live bait is always a great option for walleye and sauger. These can be fished alone, under cigar- or pencil-style fixed or slip floats with split-shot weights. A drop-shot rig is a great way to fish deep structures where you mark fish and bait on sonar. Slip-sinker rigs are most popular, as these allow anglers to stay on the move and cover an area with live baits. Walking-style slip sinkers, bottom bouncers, and no-snag sinkers are best for slowly dragging the bait along with a trolling motor. They maintain bottom contact in rocky areas and places with lots of woody snags. Rattling weights are great for attracting more fish in rivers and stained-water lakes. The use of Carolina-rig brass clackers in conjunction with glass beads, sliding and banging together on the main line between the weight and swivel, adds even more noise to the package.

Brightly colored line floats and painted floating jigheads are often used at the business end of the leader to keep the bait up out of snags and draw the attention of fish at distance. Most fishing calls for a rig with one or two short-shank live-bait hooks or mid-length shank baithold-er-style hooks from #8 to #1 in colored nickel or fluorescent paint. Top bait choices include lively medium-sized shiner and fathead minnows, or chubs, darters, small shad, or other local baitfish from about 2-1/2 to 6 inches. Night crawlers, crayfish, and of course the traditional top live bait of the north, leeches, are also great when you can find them. Dad and I have caught more walleyes on leeches throughout the United

States and Canada than on any other live baits. Amber landed a giant Ontario trophy walleye over 9 pounds, using a leech on Pipestone Lake. But they work great across Southern Illinois too. If you don't catch your own, leeches are not always available from local bait shops here though. It's best to call around, but it can be worth a little effort.

Probably the best way to fish live bait though is to combine your choices with artificial lures. Painted jigs are great for fishing any of the live baits listed here for walleyes and sauger. And I caught my biggest sauger a number of years ago on the Mississippi River with a jig-and-minnow combo. Spinner rigs are another top choice. This is a modified slip-sinker live-bait rig. It features additional beads and a clevis on the main line with a spinner blade attached, which adds vibration, noise, color, and flash to the package. Anglers use various live and dead baitfish, worms, and of course leeches on spinner rigs. These two methods are common for marble eyes and gravel lizards. But what isn't common is the use of a spinnerbait. Heavy, safety-pin-style spinnerbaits can be great, especially for bigger specimens. Using single-blade and multiblade styles, we've dragged and slow-rolled these along bottom with great success. Use a skirt in dirty water to add bulk and color, but a naked spinnerbait usually works better when it's clear. The bigger blades attract and draw fish in to the lively offering wiggling on the back.

For casting and jigging, we mostly use 6-1/2- to 7-1/2-foot bait-casting and spinning rods in light to medium power ratings. As with rods for most species, moderate to moderate-fast actions are best for crankbaits, minnow plugs and other compact hard baits, with small to medium-sized treble hooks. But fast to extra-fast actions are better for jigs, plastics, single-hook spoons, and all artificial lures and live baits fished with medium- to large-sized J-style hooks, or large trebles. With these, we use quality large-spool spinning reels and compact low-profile bait casters. For trolling, however, choose 7- to 8-1/2-foot trolling-style bait-casting rods with softer moderate actions in a medium power rating. Match these with the larger, round trolling-style bait-casting reels in line-counter models. These allow anglers to place baits at specific lengths of line behind the boat, to get them down to precise depths. Monofilament and fluorocarbon lines chosen for various presentations generally range from 6- to 14-pound test, depending on fish size and cover. But 15- to 40-pound super braids work well too, depending on the product and line diameter. Smaller-diameter lines are usually best, as long as snags don't demand thicker stuff.

Temperate Bass

Hard-fighting temperate bass are also called the true bass. These are not related in any way to the black basses, which are actually part of the sunfish family. While most anglers refer to all black basses simply as bass, *lineside* is the common nickname for these. Temperate bass are silvery white with grayish backs, many black stripes running their length, and could never be confused with the other basses. These bruisers fight incredibly hard and are a great challenge to land. But the tactics used to catch them are surprisingly simple. That's not to say they're easy to target, because often they're not. It can take time searching with sonar and watching for surface feeding activity. But when you find them, they're likely to viciously attack whatever you throw their way. There are some offerings they like better than others, and we'll break it down.

Striped bass or stripers, hybrid stripers (also called hybrid whites or wipers), white bass, yellow bass, and hybrid yellow bass are all fast-moving and aggressive pelagic predators that favor the same types of lures and baits, but in radically different sizes. Typical lures for white bass, yellow bass, hybrid yellow bass, and smaller wipers range in size from about 2 to 5 inches and 1/8 to 3/4 ounce. Lures for stripers and big wipers however, range from around 3 all the way up to 12 inches or even a little longer and can weigh from about 1/2 to 6 ounces or more. While these are opportunistic predators that can devour anything in front of them, the bulk of their diet is highly specific. Fish, fish, and more fish is what temperate bass eat. More specifically, baitfish.

A bluegill or small bass haphazardly swimming in open water could end up on the menu. But stripers, whites, yellows, and hybrids eat oily schooling pelagic baitfish almost exclusively, in most waters. As such, fishing bottom presentations designed to mimic crayfish is practically a waste of time. So is frog fishing in shallow vegetation. You get the idea. Still, jig fishing is bad and good for these fish. Dragging a brown-and-orange pitchin jig with a chunk trailer through stumps is unlikely to produce. But a white-and-silver swim jig and paddletail shad trailer is likely to get crushed, with long casts in open water. The same lures and baits produce all year round.

Floating and suspending crankbaits and minnow baits and sinking lipless crankbaits are excellent choices, especially in baitfish color patterns. Topwater lures that mimic baitfish like buzzbaits, propbaits, stickbaits, wakebaits, chuggers, and walking-style cigar plugs all produce well. Flashy casting and jigging spoons and vibrating blade baits are top choices too. Classic and hybrid swimbaits, as well as swim jigs paired with soft jerkbaits, paddletail shads, fork-tail shiners, curly-tail grubs, and other baitfish imitating soft plastic swimbaits can be dynamite.

Inline spinners are productive, as are all styles of safety-pin spinnerbaits. But the long-arm multiblade style works best, with 3, 4, or 5 willow leaf or hatchet blades. These lures closely mimic schools of the very baitfish that temperate bass hunt most often. The other lure that accomplishes this in a different way is the Alabama rig. It's a classic for whites and stripes. We cast, jig, and troll for all these fish. Power trolling with the outboard is the best way to cover a lot of water when searching them out. But once you find a school or two, you can often stay close to them with the trolling motor. And then it's common to experience multiple periods of red-hot fishing with casting and jigging techniques.

Lively baitfish in similar sizes to the lures can work great too. Gizzard and threadfin shad, mooneye, and skipjack herring are top choices. Silversides, all shiners, fathead minnows, darters, chubs and other similar baitfish all work well. These are typically slow trolled behind the boat on flat lines or off downriggers. They're also trolled off to the sides of the boat with planer boards or mast and ski systems. A simple egg sinker on the main line, with a bead, swivel, and hook, can be effective for still fishing or slow trolling suspended fish in open water.

Hook sizes range widely from #6 to 1/0 for white bass, yellow bass, and hybrids, or #2 to 8/0 for striped bass and hybrid stripers. Short-shank live bait, mid-shank baitholder, octopus, and long-shank O'Shaughnessy J hooks, wide- and extra-wide-gap standard or octopus circle hooks, or heavy-wire trebles are all productive, depending on lure or bait type and tackle selection. Fixed and slip floats, as well as balloons, are sometimes used to slow troll baitfish near the surface. These also work to cast baits to weed lines, laydowns, standing timber, boulders, riprap, docks, bridges, and other cover that borders open water. Trolling cranks and plugs for temperate bass is similar to how we troll for muskies. Speeds with live baits are slower, often pulled with the trolling motor. But artificial lures take these fish well too, with outboard motors and power-trolling tactics from 2 to about 4-1/2 miles per hour.

When casting and jigging artificials for temperate bass, fast- and extra-fast-action rods are preferred for almost all lures. The only exception being smaller cranks and plugs with lighter wire treble hooks for whites and yellows. For striper and wiper fishing with lures, and when using lures with J hooks for white and yellow bass, faster actions help greatly with hook setting. For smaller cranks and plugs with small trebles, a crankin stick is better with its softer action. The same is true for trolling rods. These should have moderate to moderate fast actions. This provides more flex with smaller trebles, and also allows for proper shock absorption on giant hard-fighting stripes caught at speed while power trolling.

Compact low-profile bait-casting reels and large-spool spinning reels are preferred for casting and jigging. While round-line counter bait casters are best on trolling rods. For yellow and white bass and their hybrids, use monofilament or fluorocarbon lines testing 6- to 10-pound test, or super braids from 15 to 30 pounds. For striped bass and hybrid stripers, opt for mono or fluoro from 12- to 30-pound test or select braided lines from 40- to 100-pound test, depending on fish size and cover present. In clear waters, adding a fluorocarbon leader to the business end of heavy braids can elicit more strikes.

Trout

Once in a while you'll see a Southern Illinois angler fly fishing for panfish. In freshwater, fly fishing is most popular with trout enthusiasts. And a variety of flies are used in our trout waters with success. However, most anglers targeting trout in Southern Illinois waters use conventional tackle far more frequently. In my younger days, I did a lot of fly fishing for a variety of species. But I rarely pick up one of my fly rods anymore, even for trout. In most cases, conventional tackle is more effective. Especially in this part of the country, anglers using conventional gear will almost always outfish fly anglers, in both numbers and size of all species of trout and other fish. Small casting and jigging spoons and vibrating blade baits take trout in 1/32- to 1/4-ounce sizes. Top conventional artificial lures include small-hair, tinsel, and marabou jigs from 1/64 to 1/8 ounce.

These materials usually outproduce plastics. But plain jigheads matched with various soft lures can sometimes be hot too. Top choices include panfish shads with the needle tail and other jerkbait-style minnow-imitating soft plastics. The second-best option is usually a hellgrammite or other small insect imitator. Occasionally, a trout is taken on a small safety-pin spinnerbait, beetle spin, or other jig spinner. But this option pales in comparison to the inline. Inline spinners are deadly trout lures, probably as effective as the hair jigs. These skinny and compact lures are more attractive to trout than the bulkier safety-pin or jig-spinner styles. Trout hammer inlines, from about 1/32 to 1/8 ounce, although a bigger version is sometimes effective. Squirting a little trout attractant onto artificial lures can increase bites, especially with slower retrieves.

Jerkbaits (minnow baits) are effective for trout from about 2 to 4-1/2 inches, and are often one of the best choices for big specimens. A shorter and fatter crankbait can sometimes produce, but nothing in comparison to a slender minnow plug. Sometimes floating models are best. But suspending jerks get the nod most of the time in both shallow and deep-diving models. Action is usually slower, but these lures can and often do produce the biggest trout of the year for anglers willing to put in a little more work for fewer fish. This is especially the case in waters like Devil's Kitchen that tend to hold bigger trout. The larger and deeper waters often hold the biggest trout in Southern Illinois, because these fish have more places to escape to. While trout fishing is often easier and faster in the smaller and shallower lakes, the increased challenge in bigger and deeper waters means greater trophy potential. Many fish will avoid capture by anglers, and in the cold depths, they can live through the heat of summer and keep growing in size. Jerkbaits are a great option for them, and one of the best producers of trophy SI trout.

Fish eggs, trout nuggets, and scented dough baits are consistent options. The dough versions seem to work best most of the time. Molding these into little balls, they're designed to look like fish eggs, a preferred food of trout. They put off a scent trail in the water that trout find irresistible. While tiny single hooks are most often used with egg baits, tiny trebles are best for nuggets and dough baits, as these will break down in the water. And like stink baits for catfish, the treble hooks hold the bait in place on the hook. Corn will take trout at times. Just one or two kernels on a tiny single live bait hook is best. And small marshmallows will tempt trout sometimes too.

Worms are a good live-bait option for trout year-round. Smaller worms like red wigglers are good for numbers of fish. But some of the biggest rainbows ever taken in Southern Illinois have come on large, whole, lively night crawlers. Live shiner and fathead minnows in small and medium sizes will take trout. And trout will also eat crickets, grasshoppers, beetles, and a variety of insects, as well as crayfish in small sizes too. Live baits, eggs, dough baits, and similar offerings can be presented under small, thin cigar- and pencil-style floats. But most of the time, presenting these on bottom with just a small hook and split shot is most effective. Hook sizes range from #14 to #6 in long-shank Aberdeen-style, mid-shank baitholder, or short-shank live-bait J hooks, or #18 to #10 size trebles.

Monofilament and fluorocarbon lines from 2- to 6-pound test are best for most trout fishing, especially in clear water. Lighter lines are not only less visible but allow better action with many small lures. Some lures can be fished on thicker stuff. But offerings like marabou jigs and skinny-tail soft plastics really spring to life on wispy lines of 4-pound test or less. For other options like spinners and minnow baits, you can get away with 8-pound test, which is beneficial in waters with big trout. For equipment, it's primarily ultralight to light power spinning rods from 6 to 7 feet, with a moderate-fast to fast action depending on presentation. These are mostly rated to cast lures from 1/32 to 1/4 ounce. Match with small spinning reels, with large spools that decrease line memory and reduce tangling.

Rough Fish

There are many different rough fish species to target, and we'll go over top tactics for a few of the most popular species in Southern Illinois. Anglers who choose to chase rough fish know just how productive this can be. Many shun the idea, feeling that these fish are beneath them. But while these are not listed as game fish, many are as sporting as most game-fish species. Rough fish can be challenging, yet we often catch high numbers of them. But these are primarily big fish that put up an incredible fight that any angler can appreciate. Yes, many "trash fish" will fight harder than most game fish swimming the same waters. And in other parts of the world, they are revered and highly respected.

I've never guided North American–born anglers just for carp, and neither have any of my pro staff guides running charter trips for my company. Still, we occasionally get them for our clients. Most of the trips I run personally now are for bass and muskies. But I've guided charters when we've chased carp after catching other species with different tactics. And even on days when we've quickly limited out on crappies, bluegills or catfish, and caught big largemouth bass, walleye, or whites, carp are often the highlight of the day because they fight so hard, for so long. Like hooking up a striper on a trout rod, it's going to be exhilarating.

Notice that I said I've never had North American–born anglers book a carp trip. I've guided anglers from Europe for muskies. Because pike are a popular species there, some European anglers dream of catching the mighty muskie. I've guided these people for carp as well. That makes for an interesting trip, trying to do muskies and carp on the same day. But while rare, I have run carp-only trips over the years. The first booking was a pair of European-born cousins staying in St. Louis that wanted to fish exclusively for carp on a full-day trip. Even though we spend little time on this species, and I told them as much, we still managed to land some big carp, and they came back again.

Carp are somewhat shy and line sensitive. If heavy line brushes up against their body, they'll often bolt out of the area. If using monofilament, it needs to be light. Think 8- to 12-pound test, which makes landing these monsters difficult. Soft braided lines can be a better option. Super braids from 15- to 30-pound test work well, especially in a dark green color that blends in with the lake floor. We use the smallest sliding sinker on the main line as possible, with a single clear bead to protect the knot and a small black swivel. Weights should be plain lead or painted black, green, or brown.

The 12- to 24-inch leader is light mono or braid. It's connected to a small, single short-shank live-bait or mid-shank baitholder J hook. This is bronze or black in finish, from #10 to #2, depending on fish and bait size. Light-power spinning rods of 7 feet or longer with moderate-fast actions work best for us. And we match these with medium-size reels with large spools. Night crawlers, red wiggler worms, corn kernels, and carp dough baits are our favorites. But small crayfish, grubs, crickets, and other insects work well too. We've caught most of our big carp in Southern Illinois's larger lakes and rivers, the Mississippi being the favorite.

Gar are tough customers that we usually target with 7- to 8-foot medium- to heavy-power bait-casting and spinning rods. These have moderate-fast to ultrafast actions. We choose quality large-spool reels, with 17- to 25-pound monofilament or fluorocarbon or 50- to 80-pound test braid. We've had our best success on the Mississippi River and its tributaries. Two different methods work great for us to catch lots of big gar. The first method is using a catfish-style slip-sinker rig to bottom fish with live and dead baits. Top choices include smaller shad, herring, sunfish, shiners, fatheads, and other baitfish. Worms also produce very well, especially big fat night crawlers that gar happily swallow right down. Crayfish are another good live bait.

The second method is with lures. Gar love to hit a variety. Notice I said hit. The mouths of gar are so hard and boney that it makes hook setting extremely difficult. The spinnerbait has drawn more gar strikes for us than any other lure, in sizes ranging from 1/4 to 1 ounce. But it's not the best for hooking, with its large heavy J hooks. The best way to get them is to add a long but thinner wire spinnerbait trailer hook. If the main hook is a 4/0 to 6/0, we'll catch more gar by adding a 1/0 to 2/0 hook on the back. Wide wobbling casting spoons with smaller hooks take gar. Floating shallow and mid-depth running crankbaits and minnow baits have also produced well, with smaller treble hooks that can penetrate a gar's mouth. The key is to use a bone-jarring hookset. We routinely set hooks hard for muskies, bass, stripers, and whites. But an even more extreme set is often required for gar. Another hot tip is to slow troll. Power trolling is usually too fast. But dragging lures behind the boat with the trolling motor catches lots of gar. When trolling is used in conjunction with a hard hookset, that's enough to slam that steel home.

Muskies and pike aren't the only fish that can be caught with the figure-eight maneuver. It's rare, but we've caught a few largemouth and white bass on the figure-eight over the years. On a media event, I was fishing offshore in the Gulf of Mexico for a variety of species with the editor of *Ontario Out of Doors* magazine, Burton Myers, and some fishing pros from Texas and Florida. We got into some big king mackerel that were chasing shallow-diving minnow plugs we were casting around schools of baitfish. While the kingfish were chasing the lures to the boat, few were actually striking. Being muskie anglers, Burton and I leaned over the tall sides of the big boat and began figure-eighting the lures. The southern boys looked at us like we'd lost our minds, until we started turning those followers into strikers! We ended up landing a number of kings from about 15 to 30 pounds, and the figure eight was getting them to commit and eat. It was a blast! Other than muskies and pike, which the technique was created for, what's the number-one kind of fish to successfully figure eight in freshwater? It's the gar. Gar will routinely follow various lures. With a stiff and heavy rod, no stretch braid, and that bone-jarring hookset, they're amazing to catch on the figure eight.

We'll also discuss drum here. Like the popular red drum and black drum in the ocean, freshwater drum grow very large and fight hard. On both lakes and rivers, we've caught lots of big drum over the years, on the same slip-sinker bottom rig we use for catfish and gar. The only exception being is that drum don't seem to take dead baits often. They prefer lively shad, shiners, and other baitfish, live night crawlers and red wigglers, and live crayfish too. We also catch plenty of drum on artificial lures. Spinnerbaits, jigs, and plastics and spoons will take them. Crankbaits, lipless cranks, and vibrating blade baits are also great choices.

Most of the time, lures ranging from 1/4- to 1-ounce sizes and 2-1/2 to 5 inches are best. Medium bait-casting and spinning reels paired with 6-1/2- to 7-1/2-foot rods with moderate to fast actions work well. Depending on cover, monofilament lines testing 10 to 20 pounds or 20- to 50-pound braids are typical. The biggest drum I ever caught, probably between 20 to 25 pounds, was taken casting a crankbait on the main river channel of the Mississippi. It was one of the most exciting battles I'd had to that point in my young life. Dad and I must have drifted at least a half mile downstream before I could get it up to the boat and he was able to slide a net under it.

So, we've discussed in detail the best lures and tackle, fishing styles, and presentation options for each individual species, like bass, or group of species, like catfish. This is the best information that anglers can use to go after their favorite fish here in lower portion of the Prairie State. Some anglers choose to dedicate most or all of their time on the water to just one species of fish, like muskies, or one group, like panfish. Many more anglers, however, enjoy catching every fish that they're able to hook up with and love everything that swims our magnificent waters. The next chapter is for those anglers. And maybe, just maybe, it can convince a few of those single-species diehards to find a little love and time for the rest.

9

Multispecies Approach

We now understand the best ways to focus individually on the various species of Southern Illinois fish. It's time to lay out tactics to catch them all, by targeting several different species at the same time. There are some residents of Southern Illinois and some traveling anglers who visit this beautiful land who are dedicated to and focused entirely on just one particular species. Most of these specialists choose to target either bass, crappies, muskies, or catfish and spend little or no time trying to catch anything else. However, there are more multispecies anglers here than those who are singularly focused. But most anglers lack a true understanding of the best ways to specifically target several different species of fish at once.

If you could pick between catching 30 of your favorite fish on a day trip or catching just 20 of your favorite fish along with another 20 fish of several other species, which would you choose? From my local guide service clients here in Southern Illinois, Southeast Missouri, and Western Kentucky to traveling anglers from all over North America and beyond, the majority choose (and have chosen for decades now) a multispecies charter if given the option. I realized many years ago that most clients call to book a trip for a particular species only because they believe that they can only target one type of fish at a time, or because they assume that a guide only charters for one species per trip. Once we let clients know that they have the choice, most are excited to have the multispecies option. And they're very enthusiastic to learn how best to do it.

Keep in mind that lure sizes vary considerably between the target species that manufacturers market them for. This is easy to see firsthand in a tackle box or on a store shelf. But think of it this way: a 10-inch plastic bass worm is tiny compared to a 10-inch muskie crankbait, even though they are the same length. Most trout flies look tiny when compared to crappie cranks, even though the two species commonly take the same offerings. A 1-ounce walleye jig is typically much shorter and more compact than a 1-ounce striper jig, and so on. While not advertised by manufacturers, the width, the depth, and the bulk of a lure are important considerations when we're talking about using in-between presentations to target multiple species at one time. Anglers can also make tweaks to any lures. Shortening and thinning the skirt of a muskie spinnerbait and removing its plastic trailer can make the lure appear much more compact. Therefore, it becomes more appropriate for walleyes, largemouth, smallmouth, or even white bass, while still fishing for muskies or stripers.

Unique Region

Southern Illinois is a great place for a true multispecies fishing approach. We've targeted many species at the same time for most of our lives here. But the first time that I fully realized how unique Southern Illinois was for this was during the filming of one of my earlier network TV shows with a crew from up north. They primarily air shows done across the northern United States and Canada. They were down for several days, hoping to fish a couple of different lakes. They planned to get a good show done for muskies. They were also hopeful that we could shoot a second show for largemouth bass or walleyes. When I told them that we could target all three species at the same time, they were very surprised. Their experience in the north country had always been that there are bass spots, muskie spots and walleye spots, that are all targeted with different lures and different tackle.

Like most of our clients and CSO pro staffers, Dad and I prefer to cast for most fish. However, jigging, trolling, and still fishing all produce well at times. And sometimes one method significantly outshines all the others. It benefits an angler to be familiar with all fishing styles and methods, even if they primarily choose one over the others. And this ups the chance to target more species together.

Walt is the best troller I've ever seen. He has mastered precision trolling to an art form. Covering every last detail, with boat, tackle, equipment, and lures, select baits are placed at exact depths, running at precise speeds, in order to target specific fish marked with sonar. During one particular week of beautiful June weather many years ago, Walt and I both had the day off from guiding and decided to get out and fish together. We had just finished up a multiple-boat trip with a group of my clients from Washington State. These were multispecies anglers, and we'd spent the charter targeting anything and everything that was biting.

We each had two guests on board. Between my boat and Walt's boat, we landed quality-sized largemouth bass, white bass, muskies, crappies, bluegills, sunfish, catfish, walleye, and one huge drum, casting, jigging, and trolling. Although we decided to primarily troll for walleyes, and while this method works well for that, we utilized tackle and lures that were different than most walleye anglers would employ. On that particular day, the bite was off for the target species, and we only caught one single walleye. What we did manage to do was to have the greatest single day of muskie fishing that either of us have ever experienced in our lives! While precision trolling, we caught an almost unimaginable 18 muskies. What's just as amazing is that 13 of the 18 fish were 40 plus inches in length, including two fat and heavy virtually identical monsters stretching the tape to 46 inches each! What's even more mind-boggling is that this isn't even all of the action that we experienced. While the fishing was red hot, for some reason the muskies weren't aggressively taking the lures deep. They were just nipping at the back hook. This made it far easier for them to shake free during the fight, especially when they got to the surface. In just about 8 hours of fishing, we hooked at least 29 muskies!

We had a few other strikes from fish that we lost before they came to the surface. Not knowing for sure what species they were, we didn't count them. Of course, some were no doubt muskies too. Since all of the fish we caught came from two main lake areas, a number of local anglers fished a little too close to us. They attempted to copy what we were doing, but to no avail. The key here is that it's highly unlikely that any muskie anglers or walleye anglers fishing with traditional muskie or walleye gear and tactics would have been able to capitalize on this special event. The angler who is prepared for anything, the one who has mastered many techniques and is willing to employ a multispecies approach, and one who has honed the senses that it takes to become one with the natural world, will be able to capitalize on more of these extremely rare occurrences in nature when fish and game cooperate so well.

Precision trolling tactics are deadly in many situations. But still, we've had spectacular days when just dropping a couple baits over the side and covering water has scored big time. Not only for muskies or walleyes but also bass, crappies, catfish, and more. I'm sure the sun, the moon, the stars, and everything else lined up just right in order for that experience to pan out. But it just goes to show that you never know what to expect out of a day on the water. Like waving a magic wand, casting for bass can produce a mess of giant crappies, drifting live baits for catfish can produce stripers, and trolling for walleyes can lead to the hottest day of muskie fishing in your life. This is especially the case when you are willing to target whatever swims our Southern Illinois waters by altering tackle and methods and employing a true multispecies approach.

Crossing Over

Many years ago, Dad and I pioneered a method of bass fishing here that we had not seen anywhere else at the time. It's still a massively underutilized method for bass, particularly in Southern Illinois, but it's one that can't be overstated. For bass, this method usually produces slower action but larger-sized fish on average, all year round. I'll break it down and explain how it ties in with crossing over for multispecies angling principals. It's produced many of our biggest bass over the years. I'm talking about mega sizing, trophy hunting with oversized baits fished on oversized tackle.

The largest bass lures, and small to even medium sizes of lures designed for stripers, muskies, pike, and saltwater bruisers, have produced many big largemouth bass for us, our guides, tournament partners, and charter clients for over two decades. Many times, when standard tactics aren't working well or only producing small bass, switching to an oversized bait is the hot ticket. Just how big are we talking? We've caught lots of largemouth bass from various Illinois waters on 8- to 11-inch saltwater crankbaits weighing 3 to 6 ounces. These are lures that we use for species like king mackerel and wahoo in the Gulf of Mexico. And one of our favorite lures for big smallmouth bass in the northern United States and southern Canada, is a 9-inch-long, 3-ounce double-bladed topwater muskie propbait!

My good friend and CSO field staff member Ken Ryder of Southern Illinois won a free fishing lodge trip to northern Minnesota's famed Lake Vermillion. And Ken invited Dad and me, along with Walt, to go along with him. Ken is one of those guys that will happily zig while everyone else zags, and he uses some unique presentations to score muskies when others fail. The primary target on this trip was muskies, on a lake known for giants. But we also focused some time on the body of water's huge smallmouth bass. That was until we realized that we could catch as many smallies on big muskie spinnerbaits and topwaters as we could on traditional bass stuff. And what a week it was. We raised some of the biggest muskies we've seen anywhere, put a 47-incher in the net, and caught some pike, perch, and walleyes. We also flat out put the smackdown on enough big brown bass that it would be worth the long drive anytime to just throw muskie lures for bronzebacks!

After many years of bass fishing in Mexico, Dad and I now rarely throw a lure that is marketed to bass anglers when we're south of the border. During my first trip down there, on a media event with Delbert Davis and Mark Shealy of the Shakespeare Pflueger Fishing Tackle group, I got more than one strange look when I opened my tackle boxes full of muskie and striper baits. Within a couple of days, however, it was apparent that lures designed for much bigger species were the hottest tickets for Mexican largemouth bass. And while not all anglers are surprised at the use of oversized baits in places like Mexico, they'd never consider them in Illinois, making a huge mistake in the process.

No doubt, some of you have likely seen us doing this kind of thing on various regional and national network television shows over the years that Dad and I have co-hosted, both here in Southern Illinois and elsewhere, with Walt Krause, Andrew Veach, and other members of our pro staff. Our favorite lures for trophy bass are the School N Shad and Hatchet Shad big-game spinnerbaits we designed for our tackle company. They're 5 to 10 inches long, with 3, 4, or 5 blades, and weigh from 1 to 4 ounces. Of course, most bass anglers would never consider throwing something that big, especially here in the Midwest. And while California big swimbait fishing has become more popular and expanded to other areas of the world over the last ten years or so, it's still not utilized in this part of the country much at all. Plus, that's a very specific kind of fishing that leaves out all other categories of lures and presentations. We're talking about a broader range of tactics here, to cover many more fishing situations. And there is no doubt that giant lures are nearly unparalleled in their ability to produce big bass.

I even co-hosted a *MidWest Outdoors* national network TV show with Larry Ladowski, in which all of the largemouth bass we caught were taken on 1-1/2-ounce, 7-1/2-inch Hatchet Shads. While these were designed for muskies and work on pike, stripers, and all big gamefish, they work equally well for all species and sizes of bass. Like they did for Larry and me that day on Kinkaid Lake, with some bonus muskies to cap off the trip. Like so many other TV shows we've done using this incredible method, this program went into tens of millions of households. And a DVD copy of the show was played for over a year in some Bass Pro Shops stores that carried these lures.

I've written articles and columns on this topic too that have appeared in *Advanced Bass Strategies, American Bass Anglers, Adventure Sports Outdoors, Bass Angler, Bass Angler's Guide, Fishing Facts, Heartland Outdoors, Mid America Outdoors, Midwest Bass Tournaments, MidWest Outdoors, River Country Outdoors, Seven 24 Outdoors, Simms Outdoors, Siouxland Outdoors,* and *Ultimate Outdoors* magazines, plus various newspapers. Despite this, and all of the TV exposure, using smaller muskie- and striper-type lures and tackle for bass is still a massively underutilized tactic in most places. And Southern Illinois is no exception. The good news? This is great for the anglers that break out of the mold, who zig while everyone else zags. I wrote my first magazine article on crossover tactics for multispecies fishing over a decade and a half ago. Since then, we've expanded these tactics for all species with incredible success. Crossing over works well for virtually all fish. Adopting a broad multispecies approach can lead to some of the most consistent and rewarding fishing of your life.

While big baits work great for largemouth bass most of the time and smallmouth bass much of the time, they even produce spotted bass, white bass, and yellows now and then. Doing the opposite, however, produces muskies, stripers, and wipers when nothing else will. Sure, big lures and natural baits are the hot ticket most of the time for these apex predators. But what do we do when they turn their noses up? Downsizing to the smallest sizes of muskie and striper lures or selecting big bass and walleye lures can be all it takes to turn around a tough day and score big time. Even average sizes of lures for bass and walleyes, considered much too small for pure stripers and muskies, are sometimes needed to get back on the board. Branching out further from this, we've used specific saltwater lures with tremendous success.

After attending a media event in South Texas where we targeted redfish, speckled trout, jacks, and other inshore species with jigs and soft plastics in the Laguna Madre, I brought back a bunch of samples of some hot new lures that wouldn't be available to the public for another year or so. These included plastics designed to perfectly mimic shrimp and crabs. They worked so well for us on Southern Illinois bass that we used them for a variety of species in Missouri and Kentucky too. When they really shined, and outproduced other plastics, was when the bite fell off and bass wouldn't strike other lure options. Crossing over works! But especially when selecting tackle and lure sizes that are applicable to many species of fish instead of just one or two.

Large and Midsized Fish

Taking this a step further is where things can really get interesting. And fun too! Most of the time we hit the water now, we'll target several species at the same time rather than chasing just one kind of fish. While cohosting a national network TV show with Greg Jones on Kinkaid Lake one year, we had a blast catching muskies and largemouth bass and targeting other similar-sized species, all on the same lures from the same spots. Greg brought the Lakemaster crew down from Minnesota, who had previously mapped major Southern Illinois lakes.

We chose rods, reels, lines, leaders, lures, and terminal tackle in the middle of the spectrum to effectively increase our chances of tying into as many different species as possible. It's easy for instance, to target a combination of largemouth bass, muskies, stripers, wipers, walleyes, gar, bowfin, and drum together at the same time with artificial lures. Of course, there are other species besides these that could potentially strike an offering intended for these fish. But these particular species mix well together, and I'll explain why.

Muskies and stripers are the giants, the apex predators at the top of the food chain. They can and do eat anything and everything. But they usually prefer larger lures. Notice I said *larger*. Largemouth bass are absolutely top predators too. And while average sizes of these fish often prefer standard sizes of typical bass lures, pay attention to the word *often*. Largemouth bass do eat a lot of big items and sometimes select for larger prey. Because of the incredibly large set of jaws on the largemouth bass, they can easily tackle much bigger prey than other species of the same size.

Walt and I took Shawn Hirst out fishing with us one day on Cedar Lake. Most times we go, striped bass are our favorite quarry. Some huge line sides have come from its depths. But on this day the stripers were nowhere to be found, and largemouth bass were more than happy to replace them. By doing what we do so often, selecting lures and tackle that would be considered smaller for stripers but bigger for largemouth, we targeted both fish and others. We scored many quality animals of one species when the others were inactive. And the giant jaws of the largemouth bass allowed us to do just that.

Walleyes are different. While the world records for largemouth bass and walleye are identical, coming in at 22 pounds and 4 ounces, the jaws of a bass are much bigger than those of the eye. What the walleye has that the largemouth doesn't, though, is a significant set of teeth. A mouth full of big sharp teeth gives the walleye an advantage with grabbing and holding onto prey. Because they can easily hang onto a big prey item, they can wait to swallow it. Unlike a largemouth bass—which swallows prey whole, taking it down in one massive gulp, with more suction power than one of those vacuum cleaners advertised on late-night TV—walleyes grab a fish, often taking their time, kill it, and then open their smaller mouths and reposition the prey to get it down easier. The teeth allow walleyes to do something that the bass can't, which gives them their own specific edge in taking down huge prey. The same is true of the similarly sized toothy bowfin.

Wipers on the other hand are just flat-out crazy. Like most temperate bass species, hybrid stripers are highly aggressive. They have to be. Being pelagic, with a high metabolism that keeps them constantly hunting, they have to eat it or lose it. Competition with other fish in their school makes them chomp first and ask questions, well, never. Wipers lack the teeth of a walleye and lack the jaw size of a largemouth bass. But their bulldog feeding style, rushing in and fighting for every scrap, makes them attack much larger items than similar-sized species would. Even without the tools the other fish have. For the most part, drum lack all of these attributes, but they make up for it in size. Gar, on the other hand, have sharp teeth and an aggressive tenacious attitude. Plus, they've got some size on top of that. On any water with two or more of these species, it becomes highly likely to catch more fish by targeting them all.

And so, we start off with the rod. It's something on the heavy end of the bass and walleye spectrum but considered light for muskie and striper fishing. Longer rods allow anglers to get away with using lighter and less powerful rods, while still landing giant fish. We select rods from about 7-1/2 to 8-1/2 feet in medium-heavy power ratings, with either moderate-fast or fast actions. Depending on brand and model, we've use rods rated for the following sizes of lures in this category: 1/4 to 1-1/4 ounces, 3/8 to 2 ounces, 1/2 to 2-1/2 ounces, 3/4 to 3 ounces, and 1 to 4 ounces.

Every one of them has worked well for crossing over to a true multispecies approach for all the species listed. Now, we're probably not going to be using a 1/4- or 3/8-ounce lure, or a 3- or 4-ounce lure, when trying to maximize the number of these species we target at one time. But even the lightest and heaviest models work well in many situations when matched with a compact low-profile bait-casting reel in a wide-spool model. For monofilament and fluorocarbon lines, it's usually 17- to 20-pound test. But we use braids more often, in sizes ranging from 40- to 65-pound test, usually finishing with a fluorocarbon leader testing 50 to 60 pounds or so.

Bulk varies considerably. Some lures are short but deep and fat, while others are long yet skinny. But generally, day in and day out, we're talking about spinnerbaits from 4 to 7-1/2 inches and 3/4 to 2 ounces, inline spinners from 4-1/2 to 7-1/2 inches and 1/2 to 1-1/2 ounces, crankbaits from 3 to 5 inches and 5/8 to 2 ounces, minnow plugs and jerkbaits from 4 to 8 inches and 1/2 to 2 ounces, topwater lures from about 4 to 6-1/2 inches and 1/2 to 1-1/2 ounces, and swim jig and plastic combos from 4-1/2 to 8 inches and 5/8 to 2 ounces. We could go through every kind of lure category, but you get the idea. Big for bass, walleyes, wipers, and gar, but small for muskies, drum, and stripers. With the right kind of equipment, you significantly up the odds on increasing the number of species targeted, and therefore the number of fish caught.

But what about those big catfish? Are we leaving them out? No way! While it can be common to catch various black and temperate bass, walleyes, muskies, and other species with traditional catfishing baits and tactics when still fishing, this occurs almost exclusively with live baits. And while flatheads and channel cats will and do sometimes attack artificial lures, they take natural baits far more often. Adding spray-on scents and scented soft plastic or pork trailers to artificial lures increases the chances of catching cats. But the best way to do it while targeting these other species is with the old drag-and-gun technique.

First employed many decades ago by bass anglers in Florida with shiners and muskie anglers in Wisconsin with suckers, it's a double-barrel approach. Anglers use the trolling motor to move along structures. We cast artificial lures at cover, while at the same time dragging a live bait or two behind the boat. Depending on depth and cover, this can be a weighted or unweighted bait, free swimming or under a float. We like to use a heavy-wire short-shank live-bait hook in a weedless model to reduce snags. We often fish it with just a little weight to keep it down, under a large cigar or pencil float or a side-planer float. Shad, sunfish, goldfish, and chubs, as well as shiners, darters, and fathead minnows all work great. Big catfish of various species will readily attack these offerings. But they also produce bass, stripers, muskies, walleyes, and rough fish too.

Small and Midsized Fish

A second fish grouping to target at the same time with artificial lures would be crappies, bluegills and other sunfish, all the black basses, white and yellow bass, drum, gar, walleyes and saugers, and trout. We go to the opposite end of the spectrum with this group. We'll choose larger panfish or trout lures as well as small bass- and walleye-type lures. This allows us to effectively target all of these species at the same time. What's more is that the smaller juvenile sizes of the big fish can be taken on these small baits too. A variety of cranks and plugs, spinners, jigs, plastics, spoons and blade baits, you name it—they will all be eaten in the right sizes. Lures typically range from about 2 to 4 inches long and weigh anywhere from about 1/16 to 3/8 ounce.

We drop down in equipment as well. We choose larger panfish rods and reels as well as the smallest bass or walleye rods and reels, and light lines that are strong enough to land decent bass on. My favorite rods for this kind of fishing range from 6 to 7 feet, in a light power rating and moderate-fast to fast action. Spinning tackle normally rules this game. But there are times when we'll use bait-casting gear too. The lightest bass and walleye bait-casting rods are usually medium-light power. Some of these have lure ratings as low as 1/8 to 3/8 ounce and can actually handle these tiny lures and produce reasonable-length casts.

The best options however are some of the inshore saltwater bait-casting rods. These light power rods are designed for small shallow-water ocean fish like speckled trout, flounder, and ladyfish. They're also used to target smaller sizes of redfish and snook when tiny lures are necessary. These special rods have lure ratings as low as 1/16 to 1/4 ounce. When matched with the smallest low-profile bait-casting reels and light lines they actually provide a reasonable package for casting cover.

One great trick that comes into play here is using the tail gunner or front runner. You can utilize a dropper to tie a small fly into the main line, 6 to 12 inches ahead of, say, a smaller spinnerbait or inline spinner. Or, by removing the back hook of a small minnow plug, crankbait, or topwater and attaching a short leader with a tiny trout jig at the other end, you get to fish two lures simultaneously. You're offering a larger lure and a tiny one to the same fish. Sometimes they bite only one. Sometimes you'll catch two fish at a time. But it works well. And at the same time, you can drag smaller-sized live baits behind the boat too, to pick up bullheads, channel cats, rough fish, or any predator previously mentioned. Consider crossing over and trying a true multispecies approach for some of the best fishing of your life, right here in the lower land of Lincoln.

10

Get Out There

A great many of the best times of my entire life have occurred on the water while fishing. I love to fish alone. Getting into nature all by myself and bathing in the silence, absorbing the perfection as it surrounds me and allowing it to soak deep into my soul is an immeasurable experience. Many people enjoy having time alone with their thoughts. But being alone in nature is something altogether different. In the natural world, one can be alone while never really being alone. You can shut out all of the stresses of everyday life while connecting with the wild and the creatures that call this place home. Alone time takes on a whole new meaning when it occurs in such a situation.

The Best of Times

Still, for me, even better times are had with family and close friends. Sharing the wild and the sport, sharing the passion with those you love cannot be compared with many things to be found in this life. And the memories last a lifetime. Southern Illinois and its surrounding regions are places where a great many families appreciate and cherish the great outdoors, as well as the sports and traditions practiced here. As a result, a significant portion of the population grows up fishing and boating and hunting and camping. When children are raised in such a way, they can develop a bond to the real world, the natural world, that cannot be replaced by the mountain of distractions of modern "civilized" society.

As children, my brother Shay and I could, and very often would, fish all day long, especially in the summer months, for bluegills and catfish and bass and trout. Mom and Dad started us out when we were just barely old enough to hold a pole, and that continued on throughout our childhood. We fished the most during the summer school break, which we eagerly looked forward to all year long. But we fished all months of the year, even through the ice on bitter days. Whether it was hot or cold or somewhere in between, our parents let us go. They gave us the freedom to become lost in the natural world, where we could truly find ourselves. It was a gift beyond measure. And I still hold on tightly to the memories as a small child of sitting in a boat with Mom and Dad and my brother. Nothing can replace that feeling of a cool breeze and the warm sun on my face, watching the water's surface glisten and flash like a million tiny mirrors, listening to the waves gently slap against the hull, catching fish, and absorbing these times into my very being, times that cannot be compared.

A little later on, we fished with our cousins Kevin and Brian Duffey and Les Trautman, who introduced us to more waters and more species. Cousin Les got me pumped about smallmouth bass at a young age. He helped me play quite a trick on one of my high school teachers, who was a dedicated bass angler. After catching a big smallmouth bass one day with Shay, Les cut the line, tied on a stick, and threw it over a tree limb, hoisting the bronzeback up to about chest height. As I stood back about ten feet or so, he had me position my hand just right as he snapped the shot. Long before digital cameras existed, the picture came out perfect. And, before letting him know that we'd pulled a fast one on him, my teacher convinced himself that I'd caught a new state record brown bass! He kept saying, over and over again, "It has to be a new record, you've got to call the state." He appreciated the joke so much that he told and retold the story many times.

Along with my brother, I caught my first striper with cousin Kevin on a trip out of town to compete in a hockey tournament. And while the trip was focused the least on it, the fishing is what remains the clearest in my mind. And this is really saying something too, considering that while Shay and I were on the ice together, he took a slapshot to the face that ended the tournament for us with a trip to the emergency room. Still, I'll never forget that fish, taken while casting a heavy underspin jig with a paddletail shad swimbait. The bone-jarring strike, the slugfest of a battle, and, best of all, sharing that experience with family. I went on to fish for bass and muskies with Kevin years later.

But it was Kevin's younger brother Brian who is credited with turning me onto muskie fishing, something that my dad wasn't all that keen on in the beginning but came to love, far more than fishing for any other species in the end.

After years of he and his mom Jean and brother Kevin fishing muskies in Wisconsin and Ontario, Brian took me along with him and a buddy of his, Tim Watson, a Mississippi River catfish guide, on my very first muskie fishing trip. It just happened to be on Kinkaid Lake. We fished hard all day long, in nasty cold weather. We didn't catch a muskie that day. In fact, we didn't catch any fish of any species. And we didn't even have a follow from a muskie. But Brian's stories of muskie fishing up north had already lit a fire in me. He gave me an old copy of the long-popular *Field & Stream* magazine, in which I'd be featured many years later. This issue contained an article about the fabled fish of a thousand casts, and I was hooked. I read everything I got my hands on about muskies and muskie fishing, talked to everyone I could, watched and rewatched every TV show about muskie fishing that I saw, attended sport shows, and went to muskie fishing seminars.

I knew going in that it was the toughest game in the sport of fishing, but I didn't care. I still fished for other species too, especially for Southern Illinois's big bass and catfish. But I spent more time chasing muskies than anything else back then. All in all, I devoted 5 years of my life to targeting the elusive muskie before I finally caught one. Yep, you read that right, 5 years! Dad and I have always fished together the most. He did fish for muskies with me quite a bit, but he threw bass lures most of the time. From way back in the old days, quite a few of my friends also fished muskies with me over the course of many years. Chris Shannon, Andrew Adank, Charlie Ewen, Nick Reynolds, and Craig Fisher are all great anglers who share my love of the outdoors. They all put in their time in pursuit of Old Esox with success. But most also spent quite a bit of time targeting other species while we fished together.

My cousin Brian, however, was different. Brian would do some bass fishing from time to time. But like me at the time, he was very devoted to the mighty muskie. The interesting thing, though, is that while he had caught a lot of pike on his trips north with his mom and brother, like me, Brian had never landed a muskie. Kevin had caught them, but Brian never did. Brian wasn't with me the day I caught my first muskie, a fat late-spring 36-incher. But thankfully Dad was. We duplicated the pattern that we used to catch it, down to the letter. As weeks went on, we caught another and then another, and the learning curve shrunk quickly. Within no time, Dad only wanted to fish for muskies, and it's still a legitimate addiction for him today, even more so than for me. Since Brian was the one to turn me onto muskie fishing, it was so cool that I was able to put him on his first muskie. It was a giant prespawn female. And Dad and Brian and I continued on, catching more muskies ever since.

On a hot summer Southern Illinois day, my nephew Cade was just five years old when he outfished his fishing pro uncle Colby and fishing pro poppy Ray. We'd barely tied up the houseboat on a spot known for a variety of species when Cade started landing bragging-sized bluegills and sunnies, one right after another. Even though he was so young, he had quite the instinct to time a proper hookset. His proud grammy Gloria looked on as Dad and I tried to keep up. But Cade caught considerably more and bigger fish than either one of us did. We'd watched a cartoon show with Cade one day, when the characters went fishing. And just like being on cue, after landing a half dozen or so nice fish, almost quoting the cartoon, he looked over at me and said, "You doing any good over there Colby?" We all still laugh about that one today.

I've made a lot of good friends from the industry over these many years of pro fishing and media work. And it's been so cool to get to share a boat with some of my idols, heroes of mine from before I shared their profession. I've been on the water with some of the best fishing pros, network television personalities, and outdoor writers in the world, legends of the sport. One of my favorite people, and probably the biggest influence over my career for the first decade, is Mark Davis, whom I've talked about earlier in this book and in quite a few of my magazine and newspaper articles and columns over the years. With Mark and through Mark, I've traveled to many of the planet's top world-class fishing destinations. But we've fished together right here in Southern Illinois too.

I've attended numerous major saltwater media events with Mark, which included offshore big-boat fishing, inshore small-boat fishing, as well as inshore wade fishing for various species. Two of my favorite stories about Mark happened during different years, but both while we'd anchored the boats and wet waded shallow saltwater flats together. One day, when we were doing well on quality-sized red drum in about waist-deep water, I suddenly became aware that I was literally surrounded by jellyfish. I yelled over to Mark, who thankfully was within earshot. I advised what these jellyfish looked like, and inquired if there was any need for concern. Always the comedian, Mark told me that I would be fine . . . as long as I didn't put one where I wouldn't put a jalapeno pepper!

I believe it was about two or three years later, we were again wet wading a different place. We were fishing pretty close to each other but wearing out the speckled trout. With a nice fish on, I wasn't exactly paying a tremendous amount of attention to my surroundings. This was when I happened to glance over my shoulder to spot a large shark heading my direction. Catching sharks is a blast. And it's commonplace to see sharks while wading the flats. In most instances, these sharks will be on the smaller side and will high-tail it out of the area once they notice you there. But sharks being sharks, they are attracted to the sounds and vibrations of struggling fish. Just like those emitted from a fish being played on a line. Some anglers wear Kevlar wading boots to protect against stingrays. I don't like these bulky things. And the easiest thing to do is simply shuffle your feet along bottom. By sliding through the sand rather than picking up and putting feet down, you avoid pinning a stingray to the bottom, which can elicit a defense sting.

This however, kicks up debris and crustaceans. Sharks and rays will get in behind an angler and feed in their mud trail. But they're usually small and not of concern. In fact, in all my years of fishing and diving, I've only had one scary encounter in the water with a big shark, which happened to be a bull that rammed into me! The one currently coming at me was also a big shark. I was about waist deep, and we were a long way from the boat. I quickly yelled over to Mark, and he spotted it just before its fins dropped below the surface as it swam in between us. Mark has much more experience than I do with large sharks. Of course, I inquired if we were okay or needed to head for shallower water. Mark said we were okay, that it was just a blacktip and not to worry. I went back to fishing. Until about fifteen minutes went by, and I looked over to see that Mark was no longer there. I looked back behind me. He was way over to the north in much shallower water! I yelled over to him again, inquiring about his locational shift. And Mark advised that he was getting away from me and that big shark!

In reality, he'd stopped catching fish and was getting more bites in shallower water, closer to a hump. But he didn't share that information.

Back then, I knew a little about blacktip sharks, but not like others. Blacktips have attacked people, but it's a rare occurrence. And they're not a species to generally get alarmed about if you're not antagonizing them. But at the time, I didn't know this. And Mark's tactics worked well to get a rise out of me. Still, I was catching lots of trout there. And I'd just hooked up on a big jack crevalle that I lost and was hoping for another. A little later on, Mark and I had waded closer together. I noticed he was singing to himself. I stopped what I was doing to listen carefully, and I heard something along the lines of, shaaaaaaaarks want Colby, du du du du du du dun, sharks want Coooooooolby . . . and on it went. He added other lyrics and sang that song off and on throughout the rest of trip—humor that was greatly appreciated by the other couple dozen anglers with us. Adventure, giant fish, and plenty of jokes, there's seldom a dull moment with Mark Davis.

I spend a lot of time fishing Southern Illinois with my wife Amber, for many different species. Being able to regularly participate in your favorite sport with a spouse that enjoys it as well is almost too great an experience for words. Before we met though, she hadn't used bait-casting equipment. Bait casting is a fine art, and it takes everyone some time to get the hang of it. Amber picked it up faster than maybe anyone else I've seen. In no time, she'd hardly even touch any other kind of equipment. On our first day out for bass, she had put down the spinning outfit for good after maybe an hour. And in no time, Amber landed her first big largemouth on bait-casting tackle. She fishes mostly for bass and muskies but has become my primary fishing partner for trophy catfish. Some anglers live to catfish, and others not so much. After all the years of fishing full-time, I like targeting big catfish at least as much as any other species in freshwater. We catch lots of Southern Illinois channel, blue, and flathead catfish, often at night. And it takes a team effort to juggle multiple rods and fight, land, and unhook numerous fish at once, especially in the dark.

After developing solid skills in Southern Illinois, Amber made her first fishing trip out of the country a while back, with me, Mom, and Dad. After we'd landed a number of quality Canadian pike, some yellow perch, and multiple 3- to 4-pound smallmouth bass, Amber caught the biggest smallie of the week that went right at 5 pounds. A couple days later, we'd been catching mostly walleyes from about 2 to 5 pounds when she hooked up to a big fish on the bottom in about 35 feet of water while targeting marble eyes in much the same way that we do on our home waters in Illinois. The beast stayed down and held its ground for quite some time. The water was crystal clear, and we saw the fish turn sideways, a good 15 feet below the surface, in a final last attempt to turn back around and dig its way to the bottom. Amber applied a little more pressure. And a few minutes later, with an eagle flying overhead and loons calling to one another across the big empty lake, Brock Clink slid the net under her biggest walleye ever. It was the second biggest eye of the trip, which went just over 9 pounds!

No Replacement

The great fishing over the years has been one heck of a ride for sure. But doing it with family and friends has made it something that words could never fully describe. If there's one thing that I could suggest here, it would be to introduce everyone that you care about to the great outdoors and to the great sport of fishing. Bring them out with you to one of the many wonderful waters of Southern Illinois. Hand them a rod and reel and teach them what you've learned. Continue on into the future, learning together, and expanding your relationships with one another as you expand your relationships with the natural world.

I've learned about waters and about fish and about the great sport of angling in each and every place I've traveled, and with many of the folks I've shared a boat or a stretch of shoreline with over these many years. But it has been on the magnificent waters of Southern Illinois where I've learned the most, sometimes alone but more often with other anglers. Because of what I've learned here, and the development of skills in this special place, I've been able to turn my passion into my life.

Southern Illinois has given me many things. Its greatest gift has been my wife, whom I met here and have fished with ever since. But the experiences of interacting with these incredible creatures in this wilderness paradise has been nothing short of a blessing. It's one that I will hold near to my heart until I draw upon my final breath. Through the immeasurable support I received from Mom and Dad, and their encouragement to chase my dreams in this special place, the inspiration that I gained from fishing here led me to devote my life to this pursuit. I could never have imagined that it would lead to what it has. That it would allow me to spend my life on the water and in the wilderness. That it would allow me the time with the fish and the animals I cherish so much. That it would take me to the places I'd fantasized about seeing when I was a child, and back again. My sincere hope is that everyone chooses to follow their dreams as I have. And I encourage everyone of Southern Illinois, and all who travel to this unique place, to use the great experiences found in the wild here to find inspiration for living, and to nurture that inspiration all the days of a long and happy life.

There is no way to replace the experiences one can have in nature, when time spent in nature is prioritized. As participants in this greatest of sports, and as residents of such a magnificent place to indulge in it, we Southern Illinois anglers have a responsibility to pass along this ancient tradition to others. And as we do, we quickly discover that not only do we light the fires of fascination, dedication, and inspiration within others, but we stoke those same fires burning deep within ourselves as well. And it's a feeling that has few equals.

Regardless of the water, the species, or the fishing styles we employ, this great of greatest pursuits refills our emotional and spiritual tanks. It gives back to us everything stolen in the hustle and bustle of the often-crazy modern world. In an overly civilized society, where false connection rules the day but real connection is a rare commodity, often with too much work and too much stress, a veritable mountain of responsibilities can pull us away from what makes us who we are, robbing us of our very humanity if we allow it. But *only* if we allow it.

Reprioritize and Renew

Concerns that have nothing to do with what humans have focused on, until just the last couple hundred years or so, sap the strength, the very will of those of us who need the wild. And guess what? Whether we all know it or not, whether we all accept the truth or choose to live outside of reality, we all need the wild. We need it, like the food in our bellies or the air in our lungs. Many lost souls today drag themselves through life, feeling that something isn't quite right, knowing that, surely, there has to be more, there has to be something to fill that big empty void.

But this doesn't have to be the case. There is a bright side to be found, a heartwarming story to be told. For by getting out into nature, into beautiful places like the woods and waters of Southern Illinois, and through participation in the modern outdoor sports blended with ancient traditions of survival, like hunting and hiking, camping and fishing, we can lose ourselves and find ourselves again. When we take it all in, soaking up the gifts of the real world, the natural world, we experience a unique and true peace, that only a deep and personal connection to the wild can provide.

Each and every time we leave civilization behind and venture into the great wilds of this amazing planet that we're lucky to call home, something may happen that we would never expect. The excitement can be difficult to contain as the wind shifts and a refreshing breeze cools us down on a hot day; as an animal emerges from the dense brush, there all the time, yet hidden from our sight; as fish come up out of the depths where we cannot see, to strike a lure we've offered with purpose and careful manipulation. These times are, simply, magical. We can feel like a kid again, on the last day of school before the glorious summer break. Like the fountain of youth, nature can rejuvenate us if we drink from it. It can refill our emotional and spiritual tanks. It can give us the strength we need to be better people as we go on through the rest of our lives and until we can get back to this place again, this place where we were truly meant to be. The woods, and the waters, and the fish await you my friends, so, get out there . . .

Index

Index

Italicized page numbers indicate figures.

COLBY SIMMS is a multiple award-winning journalist who has held staff writer, editor, or publisher positions with over two dozen periodicals during a career spanning more than two decades. He's written and sold over 3,000 works to more than 50 publications, including hundreds of cover-highlighted print magazine features. He's been the topic of many articles by other journalists, and dozens of photos of Colby and by Colby have graced the covers of magazines and newspapers across North America.

Colby is a multiple muskie, bass, and billfish tournament champion and multiple circuit championship title holder who broke numerous records and won numerous big fish awards in competition, including winning the first tournament he ever entered and winning the first championship he ever qualified for, both with his teammate and father Ray. Colby is an award-winning network television and radio personality, a sponsored professional business promoter, and a public speaker.

Colby is a record-holding guide who's operated a team of top outdoor sports pros across multiple countries. He founded one of the largest outdoor sports charter services in the lower Midwest and has personally guided many clients to once-in-a-lifetime trophy fish. With help from his mother and business manager Gloria, Colby has also designed and overseen domestic production of over 600 SKUs of his own signature sports products carried by dealers across several countries, including the world's largest outdoor sports chain retailer, Bass Pro Shops.

Colby previously worked as a private security contractor and celebrity bodyguard for international personalities of music, film, television, and business, and he served on a protection detail for President Bill Clinton. As a former law enforcement officer, along with his partner Joseph St. Clair, Colby was once widely known for exposing government corruption and winning a landmark federal court jury trial decision against a government agency that made civil rights case law.

Colby has traveled extensively with Harry Canterbury, Mark Davis, Ray Simms, and other outdoor sports legends, attending and hosting celebrity media events in multiple countries. During TV shoots, he survived a shark attack and an attempted kidnapping by members of the world's largest organized crime syndicate. Colby is a former a church leader, lifeguard, Eagle Scout, and wilderness survival instructor. He is formerly a competitive hockey player and hockey coach, a slalom skier, a martial artist holding belts in Tae Kwon Do and jujitsu, a wrestler, kickboxer, and assistant striking coach.

Colby and his wife Amber split their time between their home on the family farm in the Ozark Mountains of Missouri and their houseboat at Kinkaid Lake in Illinois, living on the water and in the wilderness. Les Winkler wrote an article about Colby's life for *The Southern*, titled "Living the Dream." Colby recently completed his first novel and is currently working on his adventure-packed autobiography that will serve as an inspirational lifestyle book, encouraging others to chase their dreams as he did and providing direction on how to do just that.

 *A Shawnee Book*